Paike Jayadeva Bhat

Galactose Regulon of Yeast

From Genetics to Systems Biology

Springer

Dr. Paike Jayadeva Bhat
Laboratory of Molecular Genetics
School of Biosciences & Bioengineering
Indian Institute of Technology
Mumbai 400 076
India
jayadeva@iitb.ac.in

ISBN 978-3-540-74014-8 e-ISBN 978-3-540-74015-5

Library of Congress Control Number: 2007937224

Cover design: WMXDesign GmbH, Heidelberg, Germany

Printed on acid-free paper

9 8 7 6 5 4 3 2 1

springer.com

Preface

Biology has captivated the imagination of researchers with diverse backgrounds as never before. For example, physicists are now exploring the origin and consequence of noise in gene expression, which appears to be important in epigenetic phenomena. Engineers are looking at biological systems from a design perspective. No doubt, conventional biologists will continue to provide insights by combining conventional approaches with high-throughput ones. These diverse efforts have resulted in the disintegration of biology into sub-disciplines. This is unavoidable because biology is inherently complex. No matter which branch of biology one studies, if the ultimate goal is to understand biology as a unitary subject, we then need to integrate these seemingly disparate aspects into a coherent whole. This is what I have attempted to do in this book.

Galactose-metabolizing enzymes are expressed in yeast upon exposure to galactose but not to glucose. This observation, first made more than a century ago, was to later become a paradigm par excellence with wide-ranging implications. The problem to be tackled here was how yeast (or any organism) adapts to changing environmental conditions. This fundamental problem continues to keep us occupied even today. Come to think of it, for survival, organisms ought to adapt. Therefore, it is not surprising that adaptation transcends every conceivable biological process that goes on in a living system. Despite considerable effort, it is only in the past few decades that we have begun to appreciate what it takes for an organism to adapt to a changing environment. It is fascinating to recapitulate the evolution of the thought processes that have brought us to our current understanding of this ubiquitous biological phenomenon. *Yeast Galactose Regulon: From Genetics to Systems Biology* encapsulates the quintessence of adaptation. Here, I have used our knowledge of how yeast adapts to galactose in a symbolic fashion to weave a common thread between wide-ranging biological themes and mechanisms as explored by conventional and contemporary approaches.

The book is divided into eight chapters. Chapter 1 summarizes the basic aspects of the yeast life cycle. I have compared this with the human life cycle to emphasize the commonality despite the evolutionary divergence. Using yeast as an example, I have conveyed that organisms are open systems, and grow at the expense of matter and energy. Knowledge of this transaction is as important as understanding the inner-workings of the cell. Chapter 2, which addresses the growth kinetics of yeast, is discussed to highlight how organisms have evolved strategies to be competitive. The fundamental concept proposed by Theodosius Dobzhansky that "Nothing in biology makes sense except in the light of evolution" is further extended to discuss the phenomenon of adaptation with specific reference to galactose utilization in yeast. This is an important transition from a generic perspective to a specific example. Chapter 3 describes identification of the genes involved in the metabolism and regulation of galactose utilization using the classical genetic approach. Chapter 4 is a continuation of classical genetic analysis to unearth the molecular interactions. These two chapters are loaded with the concepts of classical genetics, which are still being used today. They reinforce the view expressed by Victor A. McKusick that "Genetics is to biology what atomic physics is to physical sciences". Chapter 5 describes the molecular genetic experiments that have paved the way for the elucidation of molecular details at higher resolution. This is a classic example of how yeast biologists quickly embraced the growing technological breakthroughs of genetic engineering. Chapter 7 describes experiments illustrating the finer aspects of the galactose genetic switch. Finally, Chapter 8 discusses the contemporary approaches of biological analysis: genomics and systems biology. Evolutionary and applied aspects of galactose metabolism are also included in this section.

The system-centric approach followed here provided sufficient latitude to investigate the various facets of this fascinating paradigm. I have included the logic, results, and interpretations of what I think are the most important experiments. I have also included the misinterpretations of a few important experiments. These misinterpretations were not because the experiments by themselves were faulty. In some cases, the assumptions were tacitly believed to be true, while in other cases, misinterpretations were more appealing than the true alternatives that were not in conformity with the prevailing view. To a young researcher, such examples should illustrate the importance of objectivity and intuition in scientific pursuits.

Although the chapters are connected through a central theme, they can also serve as independent topics. This provides the flexibility for either downward or upward integration, depending upon one's interest and background. I have not cited the references in the text but have provided them at the end of the chapters. This is to avoid distraction from the main line of thinking. Elementary knowledge of genetics, biochemistry, and molecular genetics is all that is necessary to understand the concepts. I believe this book provides a panoramic view of how the living system can be dissected by experimental and theoretical analysis to unravel even the most minute details of biological processes that have evolved over millions of years.

September 2007 Paike Jayadeva Bhat
Mumbai

Acknowledgements

It would not have been possible to write this book without the tangible and intangible help, support, and encouragement that I received from my parents, my family, teachers, mentors, students, colleagues, peers, and friends—I earnestly thank them all. I am indebted to late Prof. P.K. Maitra of TIFR, Mumbai, for his help and encouragement during the early phase of my independent research activities at IIT Bombay, Prof. N.R. Moudgal for guiding me during the formative years of my research career and Prof. J.E. Hopper for introducing me to this fascinating topic. I thank Profs. R. Maheshwari, V. Nanjundiah, S. Durani, N.S. Punekar, K. Venkatesh, P. Balaji, S. Noronha, and Dr. R.S. Iyer for their help and valuable suggestions. I wish to acknowledge the assistance of Dr. A. Tiwari, Mr. R. Patel, Mr. Sandeep, and Ms. Swathi in drawing pictures and typing. I thank Dr. V.G. Daftary of Bharat Serums & Vaccines Ltd. Mumbai, for his concern in my research activities. I am indebted to the publishers/authors for permitting to use the data for illustration. Financial support by government organizations for doctoral, post-doctoral and my current research pursuits is gratefully acknowledged. I thank IIT Bombay for allowing me to take sabbatical and also for providing financial assistance to write this book.

Contents

Chapter 1
Introduction

1.1 An Overview

1.1.1 A General Perspective

Robert Hooke presented an account of the cells of cork to the members of the Royal Society in 1660. Although Jan Swammerdam had observed blood cells around this same time, it was only documented almost 50 years later after his death. As early as 1700, using a tiny sphere of polished glass as a microscope, which had a magnification power of 275, Antony van Leeuwenhoek observed live yeast cells as globular bodies in a drop of fermenting beer and called them "animalcules". Both Robert Hook and Antony van Leeuwenhoek had doubted the validity of the spontaneous theory, but a new approach was necessary in order to abandon the spontaneous generation theory. In 1838, while outlining the importance of the cell nucleus over dinner, Matthias Jakob Schleiden (a botanist) prompted Theodor Schwann (a zoologist) to recall observing similar structures in the notochordal cells of the tadpole. The perceived commonality of the plant and animal world, in having nucleated cells, led to the "grand unification theory" of biology, the cell theory. Identified as the common denominator of "life", the cell became its fundamental building block. However, in Schwann and Schleiden's "cell theory", cells arise spontaneously, contrary to the subsequent recognition that a new cell arises from pre-existing cells. Since the time of "grand unification" much progress has been made in dissecting cells down to the atomic constituents. Paradoxically, the fundamental essence of "life" is not fully understood.

The second half of the 18th century was a turning point. The cause and significance of heat production during fermentation was discovered by Louis Pasteur, who argued that fermentation is a process involving chemical transformation of glucose to ethanol and is intricately linked to life. Later, using the famous swan-neck experiment, Pasteur demonstrated that a cell arises only from a pre-existing cell, putting to rest the long-held belief of "spontaneous" origin. Justus von Leibig did not accept that fermentation of glucose was in any way fundamental to "life". Later, Hans and Eduard Buchner demonstrated that even yeast extract, although lacking a "living" cell, ferments glucose. Both Leibig and Pasteur were right in their own ways; "Biochemistry" was thus born, being given this name by Carl Alexander Neuberg in 1903.

P.J. Bhat, *Galactose Regulon of Yeast.*

Invariance of "form", that is, offsprings resemble the parents, is the hallmark of living beings. Over centuries, this was thought of as "blending" inheritance of the parental traits. Gregor Mendel disproved this long-held dogma in his famous pea-plant experiment, finding inheritance attributable to discrete and paired factors that do not "blend", but are passed on unchanged to progeny. Mendel's idea was much ahead of its time and it took 50 years for its importance to be realized. Independent of Mendel, Francis Galton and his student Karl Pearson observed that inheritable traits like anthropometric features did not follow the Mendelian "non-blending" or "yes-no" type of inheritance. The difference between these two schools of thought was finally resolved in the observation of Fisher that even continuously variable traits, like body height and physical strength, are "poly-genic", governed possibly by a number of independent Mendelian factors. Fundamentally, Mendel tried to explain the inheritance of "factors", and not how they determine traits, which depends on how "factors" are expressed in defining the phenotype. By the early 1900s, the existence of genes as particulate "factors" located on chromosomes in a linear fashion was clearly established.

If invariance in inheritance is due to constancy of factors, then how does one explain the diversity of life forms and the origins of species? The answer came from Charles Darwin's theory of evolution, one of the landmark scientific discoveries of all time. Darwin argued that speciation occurs by natural selection of accidental variation in genetic traits that are passed on unnoticed until such time as detected and picked up by natural selection. Revolutionary scientific ideas often have a turbulent beginning. Before Darwin's theory, the commonly held belief originally proposed by Jean Baptiste Lemark was that the traits that an organism acquires in its life time are inheritable in the progeny. According to this view, evolution does not occur by accumulated random accidents in their natural selection, but is a process of will or guidance. Although the Lemarkian theory has been abandoned for several decades, recent experimental evidence has revoked this idea all over again.

Cellular morphogenesis involving the transformation of nutrients, a process referred to as *metabolism*, inheritance of parental traits encoded in genes involving *genetics*, and origin of species due to random variations and natural selection leading to *evolution*, are the cornerstones of modern biology. These core concepts were developed as hypotheses to explain phenomenological observations, and it took almost a century to find the experimental proof for formulation of physical and chemical basis for these concepts. Ongoing efforts are focused in retracing the steps taken by evolution, with the hope that we will be able to recreate the evolutionary path taken by all forms of life.

1.1.2 An Aside on Analogy

To appreciate the significance of these fundamental concepts, let's consider a hypothetical example. For locomotion, a motor car uses a part of the energy obtained by burning fuel, and the rest is wasted as heat. Yeast cells undertake a similar process by burning fuel in the process of fermentation, harnessing the chemical energy for its metabolic needs, and then wasting the rest. There are clear instructions available in the manufacturing manual of a car company providing minute details of the manufacturing process. The sequencing of genomes has similarly resulted in complete cataloguing of the "manufacturing manual" but have not yet learned to interpret the cryptic language of the genes. We can dismantle the car and pretty much reassemble the working whole from the constituent components. Modern biology has similarly succeeded in dismantling the yeast cell or any living cell for that matter, into its constituents, but cannot reassemble the cell from its constituents. Let us further imagine that because of an accident, the design of the car is altered due to which its performance is better than its predecessor. This "new design", caused by an accident, will not be incorporated into the newer models, and neither will this information get recorded for future use. In living systems, the changes (mutations) occurring by chance or accident are not trivial events and are retained in the DNA sequences, as long as the changes are not detrimental. If in the future such changes turn out to be beneficial for the organism, they will be selected, and if not, weeded out. It is clear that a man-made machine has but a few similarities with a living cell. A living cell, a molecular ensemble capable of reproduction at the expense of matter and energy, is often compared to a computer, which performs complex tasks as programmed, but even here the analogy breaks down because it cannot reproduce itself. These comparisons only illustrate the complexities of a living cell and the problems that remain in unmasking the secrets of "life".

Where are we in our attempts to understand the interplay between the metabolism, genetics, and evolution, which dictates the myriad of living states? The problem is to understand how the phenomenon of what we call life emerges from an ensemble of inanimate molecules about which we know a great deal. The axiom that the whole is more than the some of its parts sums up the present dilemma of modern biology. In this book, I have attempted to bring out and integrate these aspects by discussing what we know about the utilization of galactose, a simple sugar, by a unicellular microbe referred to as *Saccharomyces cerevisiae*. Although the scope of this topic may appear rather narrow, the fundamental principles highlighted are applicable to all aspects of life.

References

Bowler PJ (1990) Charles Darwin: The man and his influence. Blackwell Publishers, Oxford

Delbruck M (1966) A physicist looks at biology. In: Cairns J, Stent GS, Watson JD (eds) Phage and the origins of molecular biology. Cold Spring Harbor Laboratory of Quantitative Biology, Cold Spring Harbor, New York

Fisher RA (1918) The correlation between relatives on the supposition of Mendelian inheritance. Trans R Soc 52:393–433

Forest DW (1974) Francis Galton: The life and work of a Victorian genius. Elek, London

Grene M, Dephew D (2004) The philosophy of biology. Cambridge University Press, Cambridge

Harold FM (2001) The way of the cell. Oxford University Press, Oxford

Horecker BL (1978) Yeast enzymology: retrospectives and perspectives. In: Metry B, Horekar BL, Stoppani AOM (eds) Biochemestry and genetics of yeast. Academic Press, New York, pp 1–15
Lahav N (1999) Biogenesis: theories of life's origin. Oxford University Press, Oxford
Mayr E (2004) What makes biology unique? Cambridge University Press, New York
Tudge T (2000) In Mendel's footnotes. Jonathan Cape, London

1.2 Yeast is a Eukaryotic Model Organism

1.2.1 Introduction

Living beings manifest extraordinary diversity and their classification was one of the earliest endeavors of biologists. Historically, organisms with similar anatomical and physiological features were grouped together. The earliest distinction was made between animals (which are motile and food ingesting) and plants (which are static and synthesize their own food). Later on, based on cellular organization, living beings were divided into eucaryotes (which posses a distinct nucleus) and prokaryotes (which lack this nucleus). This classification was further modified to include the variations existing between the members of eucaryotes or prokaryotes. According to this, organisms were classified into five kingdoms: Monera (prokaryotic), Animalia, Plantae, Fungi, and Protista (eucaryotic).

Classification implies that members belonging to a group are evolutionarily related. The criterion of classification has been subject to constant revision because of advances in our knowledge. Therefore, it is not surprising that organisms once thought to belong to a particular class were later shown to be a member of a different class. In the 1960s, it was realized that protein- and nucleic-acid sequences contained immense genealogical information. The current basis of classification is to incorporate genealogical information obtained through the analysis of 18S ribosomal RNA sequences. These sequences are stringently conserved and the extent of changes in the sequence can be translated into evolutionary divergence. Carl Woese, based on ribosomal sequences, grouped the members of traditional prokaryotes (which share elementary ultrastructure) into eubacteria such as *E. coli, B. subtilis*, and archaebacteria (or archaea), which represent organisms such as *H. halobium* that inhabit extreme environments. In fact, eubacteria and archaea are as distinct from one another as each is from eucaryotes. According to this classification, all organisms fall into three main domains (Fig. 1.2.1). Domains are taxonomically higher than kingdoms.

1.2.2 Model Organisms

In general, a model organism should be sufficiently well studied, and therefore can be used as a representative of a group or a kingdom of life. At the beginning of 20th century, organisms for experimental purposes were mainly chosen based on their

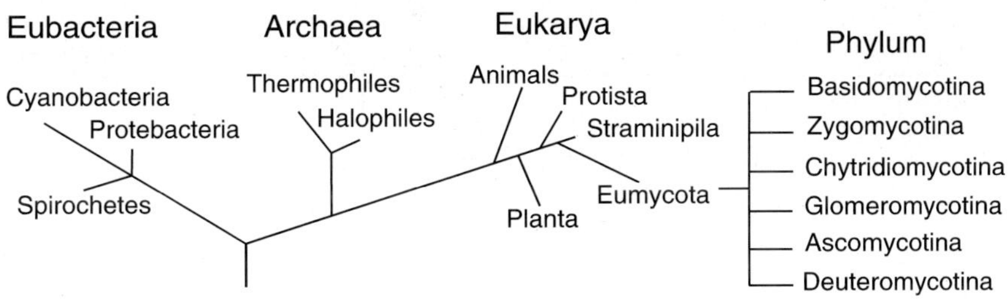

Fig. 1.2.1 Schematic illustration of phylogenetic tree that represents three domains of life forms with some examples of kingdoms in each domain. Eumycota and Straminipila (or Stramenopila) are separate kingdoms of the original group of organisms referred to as fungi. Recently these two kingdoms have been reclassified as separate domains. Organisms are classified in a hierarchical manner. The classification branches out starting with domains, kingdom, phylum, class, order, family, genus and finally, species. In some cases, variants of species are represented by infraspecies rank of *forma speciales* (Burnet 2003). The scientific name starts with the genus followed by the species

suitability for conducting experiments in genetics, biochemistry, embryology, and physiology. Accordingly, higher eukaryotes such as the fruit fly, maize, fungi, and rat were the most favored experimental organisms. Later, bacteria and their phages were extensively used to decipher the genetic basis of inheritance at the molecular level. The concept of model organisms is slowly eroding with the ability to sequence and annotate the genomes. Nevertheless, the utility of model organisms in dissecting complex biological problems is here to stay. Table 1.2.1 gives examples of the some commonly used experimental organisms.

1.2.3 Yeast

Yeasts have been the target of intense scientific investigation because of their industrial and medical importance. For example, the yeast *Candida albicans* is a common human opportunistic pathogen, and the yeast *Saccharomyces cerevisiae*, the focus of this book, is non-pathogenic and is routinely used for brewing and baking, hence the name baker's or brewer's yeast. The word yeast literally means to foam or to rise, a direct reference to fermentation. It is one of the oldest domesticated organisms and is ideal for studying various biological processes such as, development, differentiation, evolution and many others. In fact, the word enzyme, literally means "in yeast", and is a testimony for its contribution as an experimental organism for elucidating metabolism. The yeast *Saccharomyces cerevisiae* occupies a unique position as a model organism owing to the fact that its biology is understood at many levels and can serve as a reference against which other organisms can be compared. Throughout this book, *Saccharomyces cerevisiae* is referred to as yeast unless otherwise specified.

Table 1.2.1 List of model organisms

Common name	Scientific name	Domain	Cellularity
Gut bacteria	*Escherichia coli*	Eubacteria	Unicellular
Soil bacteria	*Bascillus subtilis*		
Salt bacteria	*Halobacterium halobium*	Archaea	
Thermophilic bacteria	*Methanococcus junctia*		
Bakers yeast	*Saccharomyces cerevisiae*	Eukarya	
Bread mould	*Neurospora crassa*		Multicellular
Maize	*Zea mays*		
Fruit fly	*Drossophila melanogaster*		
Wall cress	*Arabidopsis thaliana*		
Worm	*Caenorabditis elegans*		
Rats	*Rattus rattus*		
Mouse	*Mus Musculus*		

Yeast is a member of the kingdom Fungi (Fig. 1.2.1). Like animals, fungi depend on plants for their nourishment and are evolutionarily closer to animals than plants. Recent molecular evidence supports this view. A distinguishing feature of fungi is that they are either parasites (depend on a host) or saprophytes (live on decayed matter). That is, they are dependent on ready-made food and yet have been very successful. Generally, fungi are multicellular, filamentous, and multinucleate. At least 10^5 fungal species are known. It is estimated that the number of species might even exceed 1.5 million, which speaks of their evolutionary success.

Based on morphological criteria, fungi are classified into molds and yeast. Most of us are familiar with the common bread mold and the edible mushroom. Moulds are macroscopic and multicellular. The cells are cylindrical and are contiguous to one another; the structure as a whole is called a "hypha", which can branch and give rise to a mesh-like structure called a "mycelium". This cylindrical cell structure allows the organisms to penetrate the substrate. The hypha can extend at a rate of 10–20 µm/min. Unlike mold, yeast is microscopic and unicellular, i.e., a cell is an individual by itself. Because of their unicellularity, yeasts are not considered as "true" fungi, but, due to the similarity of cell wall and reproductive structures, the yeast *S. cerevisiae* is affiliated to the phylum Ascomytonia (Fig. 1.2.1). Yeast is oblate or ellipsoid in shape with two axes of equal length and a third longer one. Occasionally, the cell is nearly spherical, a form that has the smallest surface area per unit volume, thus economizing on the cell wall material needed to build an individual. Thus, yeast is a unicellular, heterotrophic eukaryotic microbe.

The natural habitat of yeast is found in fruits such as grapes, which are rich in sugar. Under laboratory conditions, yeast cells grow in a liquid or solid medium with a generation time of 90 min under nutrient-rich conditions. In a liquid medium, cells multiply until the cell density reaches a limiting value of 10^8 cells/ml. At this point, the metabolism of the cells slows down and reaches a stationary phase. If a yeast cell is placed on a solid agar medium containing nutrients, a visible colony consisting of 10^8 cells is formed within 2 days. The growth kinetics (discussed in the next chapter) varies from strain to strain and is critically dependent on growth parameters and nutritional conditions.

Table 1.2.2 Physico-chemical features of haploid and diploid yeast cells (data obtained with permission from Sherman 2002)

Parameter	Haploid cell	Diploid cell
Volume in μm^3	70	120
Wet weight (10^{-12}g)	60	80
Dry weight (10^{-12}g)	15	20
DNA (10^{-12}g)	0.017	0.034
RNA (10^{-12}g)	1.2	1.9
Protein (10^{-12}g)	6	8

Cell division in almost all organisms results in a progeny cell of equal size, meaning the parent cell grows in size and then divides into two daughter cells of equal size. In contrast, yeast divides by budding (budding yeast), which is asymmetric in that the daughter cell always is smaller than the mother cell to start with. Under favourable nutrient conditions, a yeast cell gives rise to a bud that starts growing and is followed by nuclear division and separation into two cells, a mother and a daughter cell. Yeast cells are distinguished as haploid or diploid, containing one or two copies of 16 chromosomes, respectively. Haploid and diploids cannot be morphologically distinguished under normal conditions (Table 1.2.2). However, they exhibit characteristic phenotypes depending upon the experimental conditions.

1.2.4 *Life Cycle of Haploid Yeast*

Haploid cells are of two different mating types, *MAT*a and *MAT*α (referred to as **a** or **α**) based on the genetic information present at the mating type locus (MAT locus). Haploid strains that can alternate between mating types (Fig. 1.2.2a) during mitotic growth are homothallic, while those that cannot are referred to as heterothallic. Homothallism is due to the presence of wild-type gene HO coding for endonuclease required for mating type switching while heterothallic strains lack a functional HO. Naturally occurring yeast are homothallic while laboratory strains are heterothallic, and remain as either **a** or **α** during mitotic growth.

While **a** and α are morphologically indistinguishable, they are operationally defined by their ability to mate with the cell of opposite mating type. A homothallic strain of either **a** or **α** eventually ends up as a population of diploid cells due to their ability to switch mating type during cell division. Heterothallic haploids of opposite mating type can form diploids only upon mixing either in a liquid or solid medium, thus providing an opportunity to conduct genetic analysis. Pheromone secreted by the haploids of opposite mating type (**a** cells secrete **a** factor, a peptide of 12 amino-acid residues farnesylated at the 12th Cys residue. **α** cells secrete **α** factor, a peptide of 13 amino-acid residues) interact with the complementary receptors present on the haploids of opposite mating type receptors. This interaction arrests the cells at G1 phase, and the cells become pear-shaped (these structures are referred to as *shmoos*) and fuse to form a diploid. A distinct intermediate structure resembling a dumbbell can be clearly identified during this

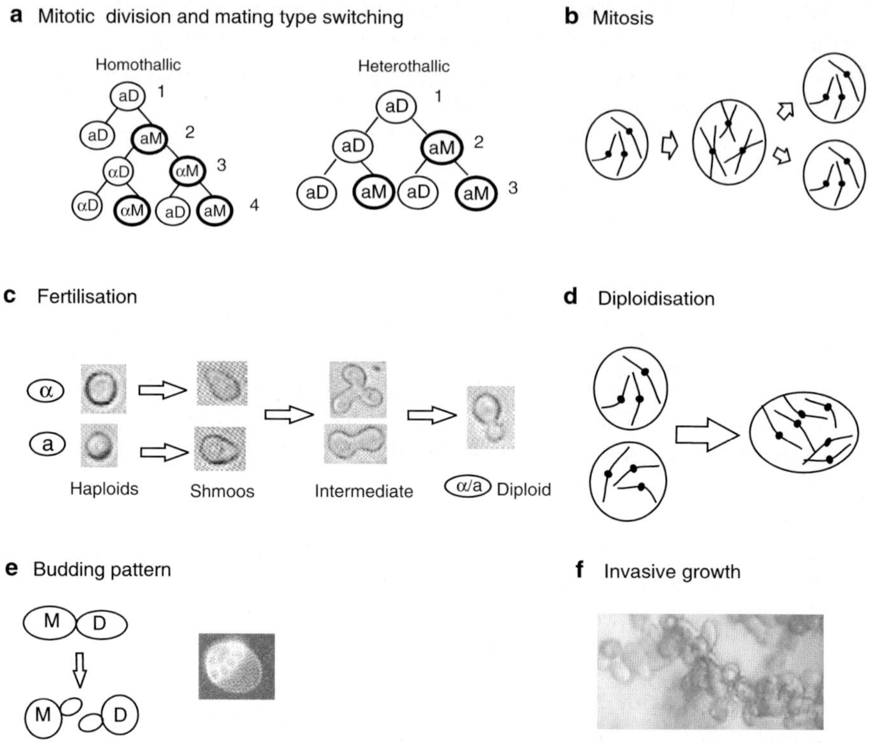

Fig. 1.2.2 Life cycle of haploid yeast. **a** Homothallic strains switch mating type during mitotic division in a predetermined fashion while heterothallic strains do not. *M* and *D* refer to mother and daughter, respectively, while **a** and α refer to the mating type. A newly formed bud (generation 1) gives rise to a daughter cell. In the next generation, when the mother cell gives rise to the second daughter, both switch the mating type. Only the mother (but not the virgin mother) is competent to undergo mating type switching. **b** Chromosome duplication followed by segregation during mitotic division. For the sake of clarity, only three chromosomes are indicated. **c** Haploids of opposite mating type fuse to form diploids. Shmoos and the dumbbell-shaped forms represent distinct intermediate stages during fertilization. Cell fusion (plasmogamy) is followed by the fusion of nuclei (karyogamy) to give rise to a diploid. Often the diploid buds at the constriction to form a clover-leaf-like structure. **d** Increase in chromosome number to twice that of the haploids is schematically indicated. **e** Haploid cells strictly follow axial budding pattern. Buds arise juxtaposed to the birth scar, and are clustered at one end of the cell. Because of the presence of chitin, bud scars appear as fluorescent spots upon calcofluor staining while birth scar is not stained (adapted with permission from Lord et al. 2002). **f** Upon depletion of glucose, haploid cells otherwise growing on the surface of agar medium as a colony, penetrate the medium. A thin section perpendicular to the agar plate on which a yeast colony was growing is shown

process (Fig. 1.2.2). Shmoos are the functional counterparts of gametes or sex cells (sperm and egg) of higher organisms. That is, in yeast, haploid differentiates into a gamete only under the influence of pheromones. This is unlike the higher eucaryotes where gametes are the product of meiotic division of a diploid germ cell (see below for more details).

Haploids of both mating type show nutrient-dependent differentiation. In a nutrient-rich liquid or solid medium, haploids strictly follow an axial mode of budding. On a solid medium, upon carbon but not nitrogen limitation, haploids invade the agar and cannot be easily washed off from the agar surface. During this differentiation, cells become elongated and switch to a bipolar budding pattern (Fig. 1.2.2).

1.2.5 Life Cycle of Diploid Yeast

Diploids do not mate. They undergo mitotic division in the presence of sufficient nutrients. In the presence of acetate, a poor carbon source, and in the complete absence of nitrogen, diploid sporulates to give rise to four haploid products encapsulated in a structure called an "ascus", which is resistant to harsh environmental conditions. In fact, sporulation is a defense mechanism to withstand nutritional limitations. On a solid medium in the presence of sufficient carbon but limited nitrogen, diploids switch from the normal bipolar budding to a unipolar mode, oval to elongated shape (Fig. 1.2.3), and cells cling to one another, giving rise to pseudohyphae.

1.2.6 Information Transfer from Parents to Descendents

Knowledge of the distribution of chromosomes during mitotic (asexual or vegetative), meiotic (reductional) cell division and diploidisation (fertilisation) is the foundation for understanding the laws of inheritance. The implications of these processes from the standpoint of evolution of genetic diversity is discussed by considering yeast and human life cycle. It is intriguing that although humans and yeast are evolutionarily separated by more than 10^9 years, at the cellular level, they show many similarities.

Mitosis: Both haploid and diploid yeast goes through mitotic division. A diploid (two sets of genetic information, 2n) yeast cell has 16 pairs of chromosomes, and the member of a pair is called homologue. During cell division, each member of the homologue gets duplicated throughout the entire length, except at the centromeric region, giving rise to an "X"-shaped structure, the arms of which are referred to as "chromatids". Thus a chromosome gives rise to two sister chromatids. Sister chromatids representing copies of each chromosome get partitioned between the newly formed bud and the mother. A haploid yeast also goes through mitotic division (Fig. 1.2.2) just the same way as the diploid, except that only 16 chromosomes participate.

A noteworthy feature of mitosis is that the daughter cell receives an identical number of chromosomes as that of the parent cell and therefore the progeny of cells derived through mitotic division are referred to as "clones". Mitotic cell division generally

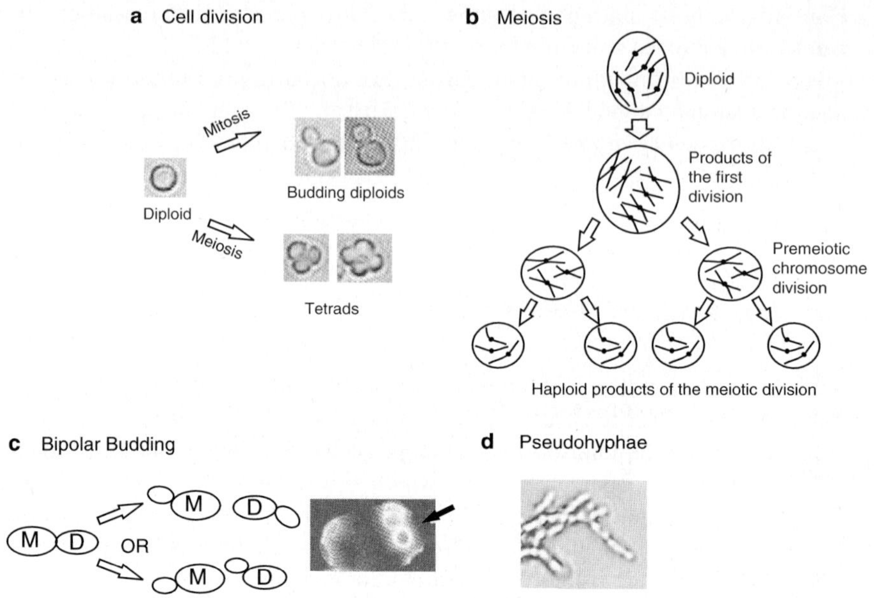

Fig. 1.2.3 Life cycle of diploid yeast. **a** Just like the haploids, diploids also divide mitotically in the presence of sufficient nutrients but sporulate by undergoing meiotic division when carbon and nitrogen are limiting. **b** Chromosome distribution during meiosis. Only three homologues are shown for clarity. **c** Under nutrient-rich conditions, diploid exhibits bipolar budding, meaning it can bud from both the birth scar end and the opposite end (adapted with permission from Lord et al. 2002). **d** In the presence of sufficient glucose but limiting nitrogen, cells become elongated, switch from normal bipolar budding to unipolar mode and the cells cling to one another, giving rise to pseudohyphal growth pattern

occurs during assimilatory growth when nutrients are in plenty and is often referred to as "vegetative reproduction". In humans, for example, the diploid zygote, meaning the first diploid cell formed by the fertilization of the haploid egg and the sperm, divides only through mitosis to give rise to a human being consisting of 10^{15} cells.

Meiosis: Only diploid cell undergoes meiosis. As mentioned before, diploid yeast (2n) sporulates by undergoing meiosis when subjected to carbon and nitrogen starvation (Fig. 1.2.3). This means that starting from a diploid cell, four haploids (each of which consists of *n* number of chromosomes) are produced. In meiosis, as with mitosis, cell division begins with just one round of chromosomal duplication, giving rise to an X-shaped structure. To produce four haploids of 16 chromosomes each, the cell must go through two successive cell divisions. The 16 pairs of duplicated chromosomes lie side by side or pair at the equatorial plate. Each such pair is referred to as a "bivalent" (recall that during mitosis, the homologues do not lie side by side but instead are distributed randomly at the equatorial plate). In the next step, the members of each pair or bivalents are pulled apart into two separate cells.

Box 1.2.1 *Chromosomes*

A DNA molecule is made up of two strands of polynucleotide chains intertwined with one another. A polynucleotide chain consists of a large number of four nucleotides A, T, G, and C (nitrogenous bases) covalently linked in a sequence that is unique to a given individual. The two polynucleotide chains align in an anti-parallel fashion and bear the complementary bases i.e., A and C pair with T and G respectively. The two strands are held by non-covalent interactions that are dictated by the complementary base pairing rule. If the sequence of bases is known in one strand, then the sequence of bases in another strand can be deduced. What is the difference between a chromosome and DNA? In the cell, DNA does not exist as a naked molecule; instead it is enwrapped by different proteins. Because of this, chromosomes exhibit some unique structural and functional features. For example, centromere is a specific region of the chromosome associated with specific proteins. The binding of the specific proteins is dictated by the sequence of the nucleotides present in that region. In yeast, this sequence is approximately 300 bp in length. The centromere attaches to the spindle fibers, which pulls the chromosomes to opposite poles during cell division. The number of base pairs in a human haploid genome is around 3×10^9. The exact sequence of the base pairs of human chromosomes has recently been determined.

Subsequently, the X-shaped structures lie randomly at the equatorial plate, just like in mitosis, and are separated into two cells resulting in four haploid cells. Haploid spores produced by meiosis are not released into the medium but are encapsulated in a single structure called an ascus (Fig. 1.2.3). These spores become metabolically active and resume their normal cell cycle only upon the availability of favorable conditions, and if not, they remain dormant.

Of the four spores, two are of **a** type and the other two are of **α** type. Upon exposure to a nutrient-rich medium, the ascus wall is degraded and the haploid spores resume mitotic growth. During this phase, mating between the opposite haploids occurs, eventually giving rise to a diploid cell population, even if the diploid is formed by fusion of a heterothallic strain. Alternatively, the haploid spores can be physically separated and grown as a separate haploid clonal population provided the haploids are heterothallic. This technique is used to carry out genetic studies, to be discussed later.

By definition, haploid products of a diploid formed by meiosis are referred to as "sex cells" or "gametes". However, as mentioned before, in yeast, gametes are produced by differentiation in the presence of pheromones. In humans, meiosis is not a response to starvation but an integral part of the life cycle. This occurs only in diploid germ cells present in gonadal tissue (testis and ovaries) whose dedicated function is to produce gametes. Unlike yeast, haploid gametes of human origin cannot initiate a life cycle of their own, meaning they do not divide by mitosis, unless they fuse to form a diploid zygote (see below for detail).

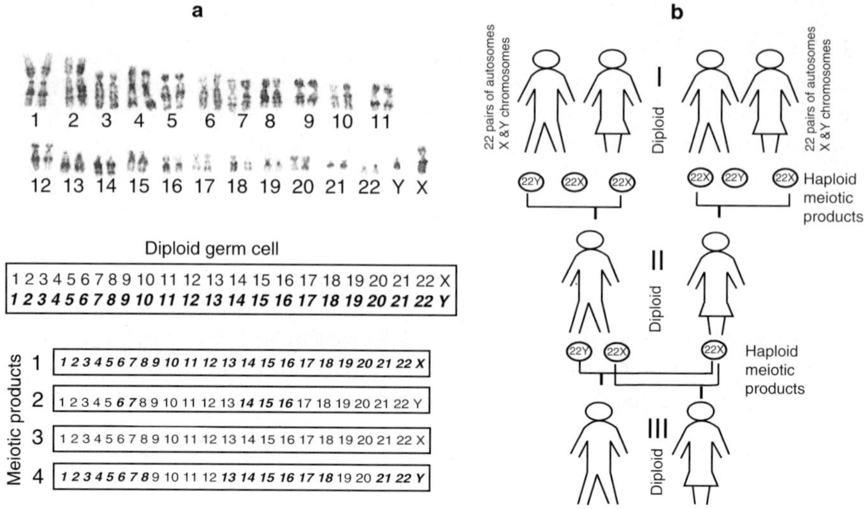

Fig. 1.2.4 Human life cycle. **a** G-banded human chromosomes arranged in homologous pairs and numbered (adapted with permission from Fincham 1994). Boxes 1 to 4 represent meiotic products with unique genetic constitution. **b** Three-generation human pedigree. In humans, the haploid and diploid states alternate. Every diploid cell contains 22 pairs of autosomes plus XY sex chromosomes in males or XX chromosomes in females. Haploid meiotic products, sperm, and egg fuse to form the diploid zygote that eventually develops into a human being. Man in family II receives 22 autosomes and Y chromosome from his father (total of 23 chromosomes) and 22 autosomes and X chromosome from mother (total of 23 chromosomes). Similarly, woman receives 22 autosomes as well as X chromosome form father and mother. The probability of this individual donating 22 autosomes and X of her father or 22 autosomes and X of her mother to the next generation is $(1/2)^{23}$. For the sake of clarity, the diagram did not consider recombination, which introduces another level of genetic diversity. This aspect will be discussed in later chapters

Box 1.2.2 *Meiosis in humans*

As the zygote divides, the descendents differentiate into different cell types, which eventually gives rise to different tissues. Thus, a group of cells gives rise to the liver, another to the brain, a third one to the gonads and so on. Cells of gonadal tissue, called germ cells, are dedicated to produce sex cells or gametes. The sex cells necessarily have to carry only half of the genetic information (i.e., 23 chromosomes and not 23 pairs of chromosomes) present in the germ cells. If this were not the case, the genetic content of successive generation would have increased exponentially and our cells would have had nothing but DNA! It is the meiotic division that maintains the constancy of chromosomes from generation to generation characteristic to a given species. Meiosis also ensures that the sex cells contain a mixture of parental chromosomes as compared to the sex cells that produced the diploid, meaning the mother's 50% genetic contribution to the egg is not the same 50% she received from her mother, or

(continued)

Box 1.2.2 (continued)

the 50% she received from her father but it is a mixture of the two. This is true with respect to the contribution coming from the father as well. The probability of egg (or sperm) receiving the same set of chromosomes donated by say one of the grandparents is $(1/2)^{23}$. In other words, during this division, a sperm or egg can have 2^{23} or 8.4×10^6 possible combinations of the 23 chromosomes, of which only one egg and a sperm would unite to give rise to an individual This explains why children of the same parents are genetically distinct from one another. Unlike in yeast, in humans, sex cells are neither encapsulated into one structure, nor are formed due to any nutritional stress. In higher animals, different gametes are produced by individuals of the opposite sex, while in higher plants, they may be produced by different structures either on the same plant or on a different plant.

Box 1.2.3 *Life span*

Two distinct processes have been identified in yeast regarding aging. One is senescence, which refers to the post-mitotic life span. Upon exhaustion of the nutrients, cells attain a stationary phase and lose viability depending on the conditions under which they attain the stationary phase. In higher organisms, post-mitotic aging is observed, for example in the brain, which depends on the survival of the post-mitotic neurons. The second phenomenon is the replicative life span. A yeast cell can give rise to a maximum of 10–15 buds and eventually the mother reaches a state of senescence. In contrast, the life span of the daughter is the same regardless of whether it is the first or the tenth daughter. This replicative life span depends only on the number of generations and not calendar time. Replicative life span is determined by counting all the daughters of a set of say 50 virgin mother cells. A somewhat similar situation exists in humans. The stem cell population of humans undergoes asymmetric divisions and the two daughters although look similar morphologically they differ in gene expression. In striated muscle, a limited supply of satellite cells is used for muscle regeneration during our life time. In patients suffering from muscle dystrophies, the life span of these cells is reduced, and cannot regenerate muscle cells.

1.2.7 Human Life Cycle

The human life cycle begins with the formation of a zygote, the first diploid formed by the union of two haploid gametes or sex cells. The zygote divides through mitosis, giving rise to the diploid cells that finally make up the human individual.

An adult individual is composed of 10^{15} cells that are genetically identical, that is, they are clones of the zygote. Although these somatic cells are genetically identical, they acquire different form and function during development and differentiation. This means that liver cells are different than say skin cells or brain cells. This differentiation occurs due to the turning on and off of genetic programs as the cells keep multiplying.

The sex cells or gametes are formed by the reductional cell division of the diploid germ cells. Reductional cell division, an integral part of the human life cycle, provides ample opportunity to shuffle the parental chromosome. This invariably results in human individuality. That is, a human individual is not a clone of either parent. Even two children of the same parents differ from one another although they may resemble each other more than an unrelated individual. This is because, the genetic constitution of the gametes issued by an individual is never genetically identical because of the mixing of the parental chromosomes (Fig. 1.2.4). Thus, the genetic identity of each individual is fixed at the time of the formation of the zygote. In fact, the purpose of meiosis is to generate as much genetic diversity as possible in the descendents. Variation between individuals is the driving force for evolution.

References

Adams A, Gottschling DE, Kaiser C, Stearns T (1997) Methods in yeast genetics. CSHL Press, Long Island

Bates AD, Maxwell A (2005) DNA topology. Oxford University Press, New York

Breitenbach M, Laun P, Heeren G, Jarolim A, Pichova A (2004) Mother cell-specific aging. In: Dickinso JR, Schweizer M (eds) The metabolism and molecular physiology of *Saccharomyces cerevisiae*. CRC Press, Boca Raton, FL, pp 20–41

Burnett J (2003) Fungal populations and species. Oxford University Press, Oxford

Calladine CR, Drew HR, Luise BF, Travers AA (2004) Understanding DNA. Academic Press

Carlile MJ, Watkinson SC (1994) The fungi. Academic Press, London

Davis RH (2003) The microbial models of molecular biology. Oxford University Press, Oxford

David M, Frazer LN (2002) Essential fungal genetics. Springer, Berlin Heidelberg New York

Dickinson JR (2004) Life cycle and morphogenesis. In: Dickinson JR, Schweizer M (eds) The metabolism and molecular physiology of *Saccharomyces cerevisiae*. CRC Press, Boca Raton, FL, pp 1–19

Fincham JRS (1994) Genetic analysis. Blackwell Science, London

Gow NAR, Gadd GM (1995) The growing fungus. Chapman Hall, London

Hans SD, Bostein KA (1986) Control of cell growth and division in *Saccharomyces cerevisiae*. CRC Crit Rev Biochem 21:153–220

Herskowitz I (1988) Life cycle of the budding yeasts *Saccharomyces cerevisiae*. Microbiol Rev 52:536–553

Kurtzman CP, Fell JW (2000) The yeasts: a taxonomic study. Elsevier, Amsterdam

Lord M, Chen T, Fujita A, Chant J (2002) Analysis of budding pattern. In: Guthrie C, Fink GR (eds) Methods in enzymology: guide to yeast genetics and molecular biology, vol. 350, Part B. Academic Press, New York, pp 131–141

Maheshwari R (2005) Fungi: experimental methods in biology. Taylor and Francis Group, LLC

Sherman F (2002) Getting started with yeast. In: Guthrie C, Fink GR (eds) Methods in enzymology: guide to yeast genetics and molecular biology, vol. 350, Part B. Academic Press, New York, pp 131–141

Strachan T, Read AP (1999) Human molecular genetics. Wiley, New York
Winderickx J, Holsbecks I, Lagatie O, Giots F, Thevelein J, Winde H (2003) From feast to famine: adaptation to nutrient availability in yeast. In: Hohmann S, Mager PWH (eds) Yeasts stress responses. Springer, Berlin Heidelberg New York, pp 305–389

1.3 A Cell as a Biochemical Entity

1.3.1 Introduction

Living beings obtain matter and energy from nutrients to make more of their kind. Nutrients are converted into a large array of biomolecules through a dynamic process involving a large number of degradative and synthetic reactions occurring simultaneously. These biochemical transactions occur with high efficiency, utmost specificity, and exquisite regulation. An intricate regulatory network at the genetic and biochemical level brings about order in the midst of apparent disorder, leading to a deterministic output, that is, cell growth and multiplication. Elucidating the biochemical features of the metabolic pathways was one of the earliest endeavors that lead to the understanding of the chemical basis of life. Here, we will discuss some of the fundamental aspects of metabolism with yeast as an example. Some features of human metabolism are also discussed to emphasize the fact that organisms differ vastly in their metabolic potential and yet, a common molecular logic pervades across the living world.

1.3.2 Chemical Constituents

Living cells contain 70–90% water by weight. The remaining 10–30% is contributed by less than 30 elements. Of these, carbon, nitrogen, oxygen, hydrogen, phosphorous, and sulphur, make up approximately 50, 20, 14, 8, 3 and 1%, respectively. These elements are referred to as macronutrients and exist mainly as organic compounds. Fe, Ca, K, S, Cl, Na, Mg, Mo, Ni, Cu are mainly found as ions in living systems, represent a minor fraction by weight, and are referred to as micronutrients. Elemental composition is fairly constant across the living world regardless of the variation in nutritional needs or the difference in the metabolic potential. This indicates that a common biochemical philosophy pervades across a myriad forms of life. Biomolecules are synthesized from precursors and organized into living entities by a teleonomic process dictated through genetic programming, which is unique to any given species. This blueprint is passed on from parents to their descendents and the cycle sustains itself without external intervention, as long as nutrients are available. Thus, living cells are highly organized self-reproducing molecular ensembles whose activities are driven by metabolism as dictated by the genetic program.

1.3.3 Macroscopic and Microscopic Aspects of Metabolism

Organisms depend mainly on solar energy, which contributes approximately 99% of the free energy of the biosphere while chemical energy constitutes a minor fraction (Fig. 1.3.1). It is estimated that 200 billion tons of carbon is assimilated in a year by utilizing 0.1% of the total solar energy available per year. Glucose, a reduced-carbon compound formed from water and CO_2 through the process of photosynthesis (Fig. 1.3.1), serves as the universal donor of chemical energy and metabolic precursor. This dual role is due to the ability of carbon to combine with itself in very many different ways, giving rise to molecules of extraordinary chemical versatility and high C–C bond energy that is almost twice that of say N–N or O–O bonds. In contrast to carbon, nitrogen-containing compounds serve as building blocks and other specialized functions but can occasionally also serve as a precursor of energy by virtue of the carbon skeleton they possess. Nitrogen available in nature in free (N_2) or oxidized (as nitrates) form has to be reduced to ammonia before it can serve as a precursor for organification (Fig. 1.3.2).

Nutrients would get depleted from the environment if they are unidirectionally withdrawn by living organisms, which evidently does not happen. This is due to a constant recycling of matter between organisms and the ecosystem. For example, carbon, nitrogen, and sulphur exist in various oxidation states due to the constant biochemical transformation occurring in living beings. Consequently, a metabolic

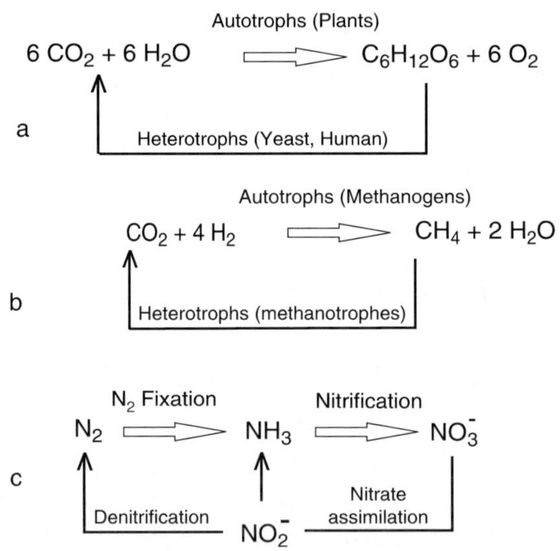

Fig. 1.3.1 Macroscopic aspects of carbon (**a** and **b**) and nitrogen (**a**) metabolism. Organisms are classified into phototropes (photosynthetic organisms like plants) and chemotropes (methanogenic Archaea) based on the use of light or chemicals as the source of energy, respectively. Organisms are also divided into autotropes (plants) and heterotropes (humans and yeast), which depend on inorganic and organic sources of carbon, respectively

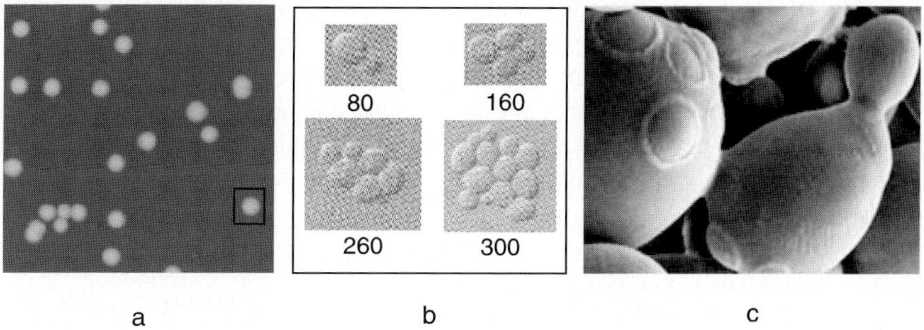

Fig. 1.3.2 Growth of yeast on solid medium with chemically defined nutrients. **a** Single cells obtained from liquid culture were spread on agar medium containing glucose and ammonium sulphate as the sole source of carbon and nitrogen. The medium also contained pantothenate, folate, inositol, niacin, amino benzoic acid pyrodoxin, riboflavin, thymin as vitamins and boron, copper, iron manganese and zinc as trace elements and potassium phosphate, magnesium sulphate sodium and calcium chloride as the salts. After spreading single cells, the plate was incubated for 2–3 days at 30°. Each colony (indicated by a *box*) represents approximately 10^8 cells. **b** Images of cell multiplication as a function of time (in minutes) after immobilization of a yeast cell on nutrient-rich agar at 30 °C. Photographs were captured from the same filed at a magnification of 20X (adapted with permission from Khron 2002). **c** Scanning electron micrograph of yeast cells showing the bud scars (adapted with permission from Angela Dunn and Mick Tuite, University of Kent, UK)

end product of one organism serves as an energy source for the other and vice versa. For example, the CO_2 produced during the oxidation of carbon compounds is the precursor for the synthesis of glucose by photosynthesis. Similarly, free nitrogen is reduced to ammonia so as to make its way for its assimilation into biomolecules. This uninterrupted (except for the human intervention) interdependence constitutes mainly cycles of carbon, nitrogen and sulphur supply in biosphere, which maintains a delicate chemical harmony (Fig. 1.3.1).

1.3.4 Biochemical Transactions

Organisms vary considerably in their minimal nutritional requirement. For example, yeast can assimilate inorganic nitrogen and sulphur available as ammonium sulphate into biomolecules, but cannot assimilate inorganic carbon. Its carbon needs are satisfied by reduced-carbon compounds such as sugars, ethanol, glycerol, etc. Like yeast, humans also cannot use inorganic carbon, but unlike yeast, humans cannot use inorganic nitrogen or sulphur and therefore depend on organic sources such amino acids.

Connectivity between the carbon and nitrogen metabolism illustrates the complementary facets of metabolism that is anabolism and catabolism. Consider the growth of yeast in the presence of glucose as the sole source of energy and carbon and NH_3 as sole source of nitrogen. In addition to this, yeast has to be provided with other nutrients such vitamins, sulphur, phosphorous salts, and trace elements, etc. (Fig. 1.3.2). Yeast obtains free energy for the synthesis of energy-rich molecules such as adenosine triphosphate by fermenting glucose into ethanol. Glucose is also converted to alpha-ketoglutarate (Fig. 1.3.3), a key metabolic intermediate that provides a carbon skeleton. Nitrogen, present in ammonia, enters the metabolic web through the conversion of alpha-ketoglutarate (derived from glucose) to glutamate. Nitrogen incorporated into glutamate finds its way into different nitrogenous molecules through a variety of biochemical reactions. For example, pyruvate can be converted to alanine in one step as oxaloacetate can be converted to aspartate. Thus, yeast can synthesize all 20 amino acids (the precursors for protein synthesis) by diverting carbon and nitrogen obtained from glucose and ammonia, respectively, (Fig. 1.3.3). Thus, catabolic and anabolic processes are intricately intertwined. The catabolic processes yield energy and precursors while anabolic processors synthesize biomolecules at the expense of the energy generated from catabolism. Essentially the catabolic pathways converge and the anabolic pathways diverge.

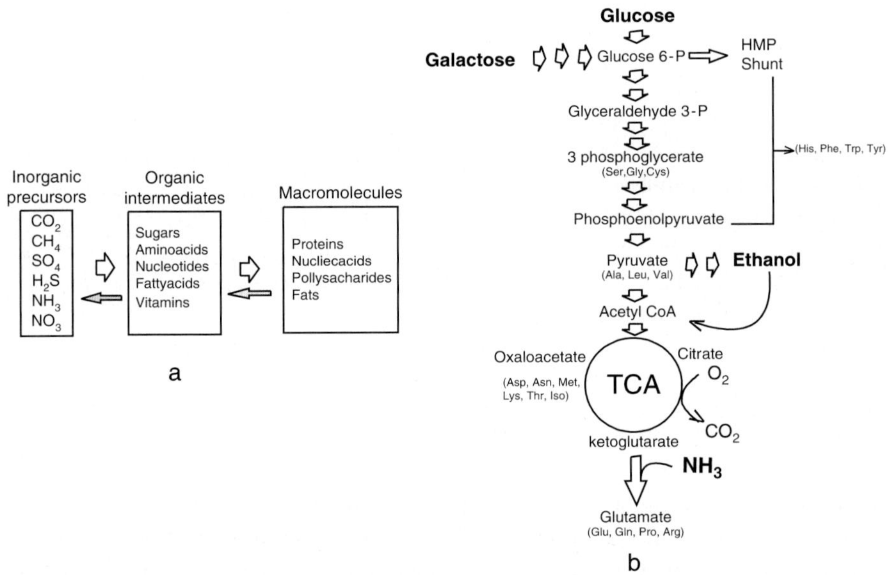

Fig. 1.3.3 Interconversion of chemical constituents and the central metabolic grid. **a** Inter-conversion of inorganic precursors into organic intermediates which in turn serve as precursors for macromolecules. *Open arrows* indicate assimilation (anabolism) while *shaded arrows* indicate degradation (catabolism). **b** Metabolic grid indicating the connectivity between carbon and nitrogen metabolism. Yeast grows in the presence of a variety of carbon and energy sources such as glucose and galactose. Ethanol, an end product of sugar catabolism, is also used as a source of carbon upon exhaustion of the fermentable sugars. Similarly, instead of ammonia, it can sustain on other nitrogen sources such as urea or amino acids. For example, starting from the carbon and nitrogen of glucose and ammonia, respectively, yeast can synthesize 20 amino acids (indicated in *brackets*) that serve as precursors for the synthesis of proteins

Box 1.3.1 *Human nutritional requirement*

Humans are fastidious in their nutritional requirement with regard to both carbon and nitrogen. In humans, glucose is indispensable as a source of energy although fatty acids present in the normal diet serve as a source of energy. Glucose metabolism not only varies from tissue to tissue but also depends on the physiological state. For example, red blood cells (RBC) and brain cells are exclusively dependent on glucose for energy. While RBCs derive energy mainly by fermentation, brain cells oxidize glucose to CO_2 and water through mitochondrial oxidation. Cells of skeletal muscles normally oxidize glucose to CO_2 and water, but during extreme exertion, glucose is fermented to lactate. On the other hand, cells of cardiac muscle exclusively convert glucose to carbon dioxide and water. Although humans have the metabolic potential to convert alpha-ketoglutarate to glutamate in the presence of ammonia, they cannot use ammonia as the source of nitrogen. In fact, ammonia produced above a certain level due the catabolism of say amino acids is toxic and is excreted as urea. Humans mainly depend on amino acids obtained from proteins for their nitrogen supply.

1.3.5 Energy Transactions

Free-living organisms such as yeast normally do not store energy. Instead, nutrients are continuously used up for propagation. In case of nutrient limitation, they switch over to a dormant state characterized by the accumulation of glycogen or threolose, which prepares yeast to remain in a dormant state for a prolonged period of time. In contrast, in higher eukaryotes such as humans, excess energy available in the diet is stored as chemical energy in the form of glycogen or triglycerides, which is mobilized depending on the physiological context. For example, when the blood glucose level falls below a threshold, fatty acids are mobilized from triglycerides, or when there is a sudden demand for muscle contraction glycogen is broken down, releasing glucose. Unlike microorganisms, they do not have the ability to sustain prolonged period of draught.

Adenosine triphosphate (ATP) is the common cellular energy currency regardless of whether a cell uses light or chemicals as the primary source of energy. The use of ATP is similar to the use of a common currency for financial transactions regardless of the source of wealth. There are other energy-rich molecules such as creatinine phosphate that serve a similar role as ATP but they are not as ubiquitous. ATP is not a storehouse of energy, but is used as a rechargeable battery. ATP links energy-yielding reactions to cellular processes that need energy input. For example, although glucose contains chemical energy, it cannot be used for muscular contraction, but ATP can. That is, no mechanism exists to transfer the energy directly from glucose to say muscle contraction. How do living beings trap chemical energy present in say glucose or fatty acids into ATP?

In man-made machines, chemical energy is converted to heat, which is eventually converted to mechanical energy. For example, cellulose, a polymer of glucose present in wood, can be oxidized by burning to generate heat from which mechanical energy can be derived. Obviously, this strategy cannot be used by living beings because they cannot tolerate wide fluctuations in temperature. Living systems also oxidize glucose to obtain energy but without increasing the temperature. They do so by trapping free energy of glucose oxidation. What is free energy? Free energy is capable of doing work at a constant temperature and pressure.

We shall focus on how chemical transformation of glucose to carbon dioxide and ethanol liberates free energy and is trapped as ATP. The release of free energy during a chemical reaction is dependent on the difference in the chemical potential between reactants and products. Chemical potential is a product of two factors: the activity coefficient dependent on the chemical structure of the compound and the molar concentration. Since the concentrations of metabolites in living systems are in the order of mM or μM range, the activity coefficient by convention is considered as 1. The standard energy state, a unique property of the molecule, relates to the free energy of a 1 M solution at atmospheric pressure at 25 °C. An increase or decrease in molar concentration changes in energy status by a logarithmic factor.

Let us consider the conversion of 1,3 bisphosphoglycerate (1,3 DPG) to 3 phosphoglycerate (3PGA), which occurs spontaneously and is one of the two steps of fermentation of glucose to ethanol where energy released is trapped as ATP. We know that spontaneous reaction releases free energy. How do we know that the above reaction occurs spontaneously? It can be demonstrated that if phosphoglycerate kinase is added to a mixture of 1 M of 1,3 DPG and 1 M 3PGA, concentration of 1,3DPG would decrease spontaneously and 3PGA will increase (remember that enzyme does not disturb the energetics but only hastens the reaction rate). The reverse reaction, however, cannot occur starting with 1 M concentrations of both, meaning that the reverse is not spontaneous. Eventually the concentrations of these two will reach an equilibrium concentration and remain unaltered. Since conversion of 1,3DPG to 3PGA occurs spontaneously, energy is released during this process. The Keq for this reaction id 1×10^{10}, which means that at equilibrium, almost all of the 1.3 DPG is converted to 3PGA (at equilibrium, the concentration of 3DPG is close to 2 M if we start with 1 M concentrations of both). The Keq of 1×10^{10} corresponds to a free energy of 11,000 cal/mole, as given by the equation $\Delta G^0 = -RT\ln Keq$. That is, if we maintain 1 M concentrations of 1,3DPG and 3PGA, and allow 1 M of 3PGA to be formed, then the energy liberated will be 11,000 cal/mole. As a corollary, 11,000 calories of energy have to be provided to convert one mole of 3PGA to 1,3 DPG starting form the standard state. As the reaction proceeds from the standard state towards equilibrium, free energy keeps decreasing at every instant of time as the concentration of 1,3 DPG approaches the equilibrium and eventually the free energy available will be zero. Therefore, it follows that to extract more free energy form spontaneous reactions, the concentrations of reactants should be kept as far from equilibrium concentrations as possible.

Keq and ΔG^0 is characteristic of a given reaction under specified conditions. Comparison of ΔG^0 or Keq between any two reactions gives an idea about the difference in the ability of different reactions to liberate energy, or its spontaneity or the directionality under standard condition. For example, under the standard condition, ATP hydrolysis to ADP and Pi gives 7,000 cal/mole and Keq is 2×10^5.

It is to be noted that the free energy obtained under intracellular conditions is not equal to ΔG^0, but ΔG (Fig. 1.3.4). The living cell has taken advantage of the thermodynamic concepts to maintain a living state. Let us clarify the above by considering an example. In yeast, ATP, ADP, (Adenosine diphosphate) and Pi (inorganic phosphate) concentrations are in the range of 2.25, 0.25, and 1.65 mM, respectively. Under these intracellular conditions, the mass action ratio (τ) is far away from the equilibrium ratio and therefore ATP is spontaneously hydrolyzed to provide energy. How does the cell replenish the ATP from ADP and Pi? This is achieved for example, by the conversion of 1,3DPG to 3PGA. Intracellular concentration of 1,3DPG and 3PGA is of the order of 3.5 and 1.5 mM and under these concentrations, the conversion of 1,3 DPG to 3PGA is spontaneous and therefore energy is released during its conversion. In fact, the uphill reaction of ATP synthesis at the concentration mentioned above is driven by the downhill conversion of 1,3 DPG to 3PGA. Here, the energy release and capture are mechanistically coupled, meaning that the same enzyme molecule brings about both the transformations.

In principle, conversion of 1,3 DPG to 3PGA and synthesis of ATP from ADP +Pi would come to an equilibrium as the reaction proceeds. That is, 3PGA will tend to reach 10^{10} times more than 1,3 DPG (Keq 10^{10}) and similarly the product of concentration of ADP and Pi will tend to reach 2×10^5 (Keq for the reaction)

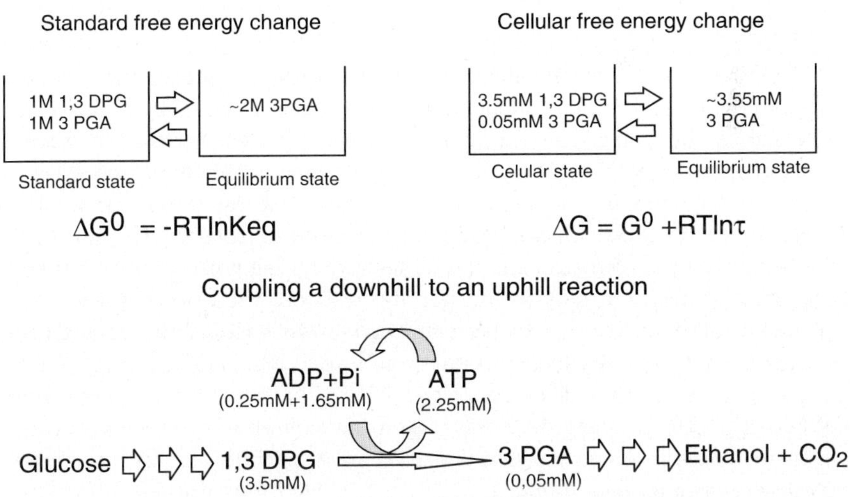

Fig. 1.3.4 Thermodynamics of metabolism. Keq refers to the ratio of the products of the concentrations of the products to the products of the concentration of the reactants at equilibrium. τ refers to the mass action ratio of the products to the reactants

times more than ATP. Living cells cannot afford to reach this state because no net energy would be available once this state is reached. One way to circumvent this problem is to decrease the ATP concentration and increase the ADP and Pi concentrations to drive the synthesis of ATP. However, this is not a viable idea since free energy should be continually available for the living state to be maintained. The other possibility is to ensure that the 1,3 DPG and 3PGA concentrations never reach equilibrium concentrations, or the 3PGA concentrations are not allowed to increase, or both.

How does a living cell achieve this? This is achieved by constantly supplying glucose and removing ethanol continuously, such that the concentration of 1,3 DPG and 3GPA is always 3.5 and 0.05 mM, respectively. Remember, what the cell has achieved is a steady dynamic equilibrium state, different from the thermodynamic equilibrium state. Yeast consumes glucose continuously and never allows ethanol to accumulate above a certain level. That is, yeast is an open steady-state system where matter and energy are exchanged. By now it must have been clear that the reaction can be made to go in either direction by changing the concentration of the reactants and products accordingly. So far we discussed the use of chemical potential for harnessing free energy. Besides the chemical potential, cells also use redox potential or concentration gradients to generate ATP.

Box 1.3.2 *Free energy and spontaneity*

Any process that occurs spontaneously releases energy. For example, the flow of water from a reservoir is spontaneous and releases free energy. This energy released can be trapped, provided a mechanism exists. Spontaneity should not be confused with the rate of the reaction. For example, although conversion of glucose to ethanol is spontaneous, glucose is stable at room temperature. Outside the living cell we need to ignite to convert glucose to CO_2 and water. In living systems, the reactions are catalyzed by enzymes. Remember that enzymes are required only to catalyze the reactions. That is, they provide a pathway for the substrates to get converted to the product by reducing the energy barriers there by increasing the rate several hundred-fold. Enzymes do not alter the energetics. The efficiency of energy conservation is never 100%. When efficiency reaches 100%, the system is at equilibrium and no net energy will be available. To ensure that this does not happen during energy conversions, a certain amount of energy is wasted as heat. This wastage is the price we pay for living. This is what Pasteur meant when he said that the heat produced during fermentation is to the benefit of the organism that ferments.

Box 1.3.3 *Thermodynamics and biology*

The law of conservation and transformation of energy was discovered during the course of investigations of organisms. R.J. Meyer observed that the color of the venous blood in humans who live in a tropical climate is quite similar to arterial blood. He inferred from this that when the temperature of the environment is raised, a lower expenditure of energy is necessary to maintain a constant body temperature. In 1842, Mayer conjectured that energy is conserved and also estimated the mechanical equivalent of heat based on thermal properties of gases. It is less well known that H. Helmholtz, discoverer of the first law of thermodynamics, also started his investigation based on observations made from biological system.

Box 1.3.4 *Forms of energy*

Energy can be regarded as occurring in two major forms: thermal and non-thermal. Inaddition, non-thermal energy occurs in several alternate forms, these being chemical, electrical, kinetic, mechanical, potential,and radiant energy. All forms of energy are inter-convertible. However, although thermal energy cannot be quantitatively converted into non-thermal energy, all forms of non-thermal energy can be quantitatively converted into thermal energy. For this reason, the quantity of energy has been historically expressed in thermal units.

References

Battley EH (1987) Energetics of microbial growth. Wiley, New York

Bullock C (2000) The archaea: a biochemical perspective. Biochem Mol Biol Educ 28:186–191

Cartledge TG, Drijver-de Haas JS, Jenkins RO, Middelbeek EJ (1992) In: Cartledge TG (ed) In vitro cultivation of microorganisms. Butterworth-Heinemann Ltd., Oxford, pp 80–106

Gilbert HF (2000) Basic concepts in biochemistry. McGraw-Hill, New York

Harold FM (2001) The way of the cell. Oxford University Press, New York

Haynie DT (2001) Biological thermodynamics. Cambridge University Press, Cambridge

Khron SJ (2002) Digital time-lapse microscopy of yeast cell growth. In: Guthrie C, Fink GR (eds) Methods in enzymology. Guide to yeast genetics and molecular and cell biology, vol. 350, Part C. Academic Press, New York, pp 3–41

Segel IH (1976) Biochemical calculations. Wiley, New York

Simpkins I (1993) General principles of biochemical investigations in practical biochemistry. In: Wilson K, Walker J (eds) Practical biochemistry. Cambridge University Press, New York, pp 1–79

Vol'kenshtein MV (1970) Molecules and life: an introduction to molecular biology. Plenum Press, New York

Wood WB, Wilson JH, Benbow RM, Hood LE (1981) Biochemistry: A problems approach. Benjamin/Cummings Publishing Company, Inc., Menlo Park

Chapter 2
Adaptation to Environment

2.1 Growth and Multiplication

2.1.1 Introduction

Of the many carbon sources, yeast prefers glucose. In the presence of glucose, enzymes of the galactose metabolic pathway are not expressed. Even a disaccharide containing glucose moiety such as sucrose, is not utilized until all the available free glucose is completely consumed. Yeast maintains a strict hierarchy in terms of sugar utilization and glucose is at the top. Does it offer any advantage to yeast despite the free energy content between say galactose and glucose is the same? Growth is a resultant of all the biological activities of a cell. Quantitative analysis of growth provides insights into the metabolic strategies adapted by different organisms or same organisms under different experimental or physiological conditions. In this section, I shall briefly discuss the basic aspects of cell growth with a focus on what a cell or a living organism considers important for its evolutionary success.

2.1.2 Growth Kinetics

Growth rate is determined by measuring cell number or biomass as a function of time. Cell number is measured by monitoring the optical density of the cell culture, counting the cell number using a haemocytometer, or determining the viable cell count. Biomass is determined by measuring the dry weight of the cells.

An increase in cell number obviously reflects an increase in biomass. However, an increase in biomass need not necessarily be due to an increase in cell number. This can occur due to an accumulation of cell material without a concomitant increase in cell number (Fig. 2.1.1a). If the cells present at the start of the experiment are at the same stage of cell cycle, that is they are synchronized, then the increase in cell number occurs in discrete steps (Fig. 2.1.1b). For most of the growth studies, it is not required to monitor the growth of a synchronously growing population.

P.J. Bhat, *Galactose Regulon of Yeast*.

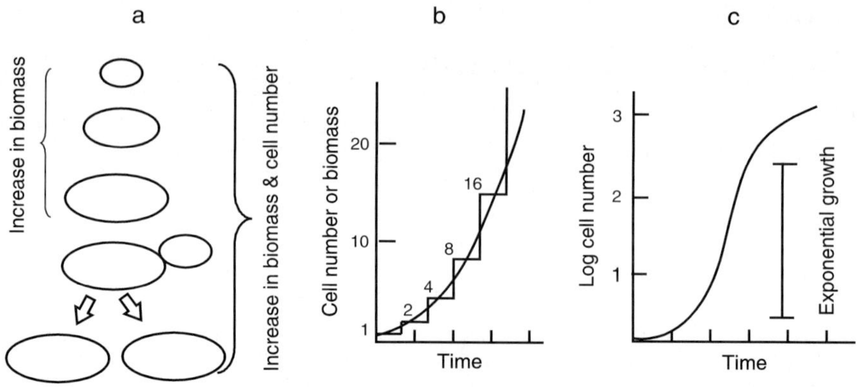

Fig. 2.1.1 Kinetics of growth: **a** The relationship between cell growth and cell multiplication. **b** The increase in cell number in a step-wise manner starting from a synchronized cell population. A curve with continuously increasing slope is obtained if cell numbers of an unsynchronized population is plotted against time. The curve also indicates the increase in biomass of a synchronized or unsynchronized cell population. **c** The growth of an unsynchronized cell population on a log scale

Unless the cell cycle is synchronized using special techniques, the initial cell population is heterogeneous with respect to the growth cycle. Because of this, cell division is a continuum, i.e., in a population of cells, at any given instant of time, cells would be at different stages of cell cycle. Therefore, if the cell number is plotted as a function of time or number of generations lapsed (Fig. 2.1.1) on arithmetic coordinates, a curve with constantly increasing slope is obtained (Fig. 2.1.1b). A similar pattern is also obtained if the biomass is plotted instead of cell number. That is, the rate of increase escalates as a function of growth. Cell multiplication of a synchronized cell population would show step-wise growth (Fig. 2.1.1b).

A typical growth profile of yeast in a batch culture is represented in Fig. 2.1.1c. The initial cell population in the inoculums is heterogeneous with respect to the physiological state and therefore different cells multiply at different rates. This initial phase of the growth profile is the lag phase, which is characterized by a slow growth rate. Once the cells adapt to the new environment, all the cells start multiplying at the same rate and the cell density increases exponentially and eventually reaches a stationary phase beyond which the cells do not multiply. Attainment of a stationary phase is due to a number of factors, such as diminishing nutrients and accumulation of metabolites. Under typical experimental conditions, yeast can grow up to a cell density of 10^8 to 10^9 cells/ml.

A fundamental parameter that describes growth is its rate. The time period required to double the cell density is called "doubling time" (t_d), which is fixed for a given organism under a given set of experimental conditions. Total cell (N_t) number at any time point during growth is proportional to the number of generations and can be calculated using the following equation, provided we know the initial cell number (N_0) and the number of generations n (n = total time/doubling time).

$$N_t = N_0 \cdot 2^n$$

How do we calculate the doubling time? The number of cells can be plotted either as a function of time or as function of generations. This representation of growth kinetics is inconvenient and needs to be transformed into a more suitable form. In the above case, the number of cells increase by geometric progression, but the parameter on the X-axis is in arithmetic progression. To convert the geometric increase in cell number to a linear form, one needs to express the above equation in a logarithmic form, which is

$$\log N_t = \log N_0 + n \log 2$$

From the above equation one can calculate *n* by experimentally determining N_0 and N_t

$$n = (\log N_t - \log N_0)/\log 2$$

If we want to calculate the number of generations per unit of time, then

$$n/t = (\log N_t - \log N_0)/t \log 2$$

n/t is called the growth rate constant k. The inverse of the growth rate constant is t_d.

$$t_d = t/n = 1/k$$

Therefore, $N_t = No\ 2^{kt}$

The increase in cell number occurs by a factor of k. The unit of k is t^{-1}. Here, the increase is considered to occur in a discrete step-wise manner, in other words, k is an average value for the population over a finite period of time, but we know growth occurs in a continuous manner. For this purpose, we need to have an instantaneous growth rate constant. This is because growth occurs even without cell division and we need to account for this as well. For this purpose, we need to consider a small time interval to get instantaneous increase in growth. That is, for a small increase in say cell mass per small increase in time.
dx/dt~cell mass existing at that instant of time.

$$dx/dt = \mu X$$

μ is the proportionality constant designated as specific growth rate constant with t^{-1} as unit

$$\mu = \text{rate of growth/amount of biomass}$$

To calculate the increase in cell mass that occurs between any two time points, cell growth occurring during small time periods will have to be added up. Mathematically, this is achieved by integrating the equation $dx/dt = \mu X$.

Thus we get $X_t = X_0.e^{\mu t}$.

This is same as the equation that we got previously and is transformed into logarithmic form

$$1n\ X_t = 1nX_0 + \mu t$$

From the above equation we can calculate

$$\mu = 1nX_t - 1nX_0/t$$

The actual increase in biomass per unit time becomes greater at each instant during exponential growth and the growth rate remains constant. Thus, a value of $\mu = 0.1\,h^{-1}$ is equivalent to a 10% increase per hour. This does not mean that the doubling of cell density would occur in 10 h, but would occur in 6.93 h. This is because the increase occurs in a continuous fashion, similar to an increase in compound interest.

μ and k are related as shown

$$2^n = (e^{0.693})^{kt}$$

$$e^{0.693k.t} = e^{\mu t}$$

$$kt0.693 = \mu t,$$

$$\mu = k\ 0.693$$

$$t_d = 0.693/\mu$$

2.1.3 *Effect of Nutrients on Growth*

Nutrient concentration affects the growth rate and the total biomass. The nutrient that limits the growth in this way is called the "limiting nutrient". For example, if yeast is grown in a medium containing a different amount of glucose keeping other nutrients unlimited, μ would keep increasing and reach a maximum (Fig. 2.1.2, panel A). The relationship between the specific growth rate and the nutrient concentration is hyperbolic (Fig. 2.1.2b). The concentration of limiting nutrient at which the μ is maximum is called as μ_{max}. Nutrient concentration at ½ μ_{max} is Ks. The relationship between μ and the substrate concentration is given by an empirical formula $\mu = [S]\ \mu_{max}\ /Ks + [S]$, given by Monod.

A linear relationship is obtained between the net biomass and the concentration of limiting nutrient over a wide concentration range. The mass of cells produced per unit of nutrient is called the "growth yield coefficient" or "yield constant", defined as $Ys = X - Xo/S$; X is the dry weight of cell (mg/liter) at the beginning of stationary

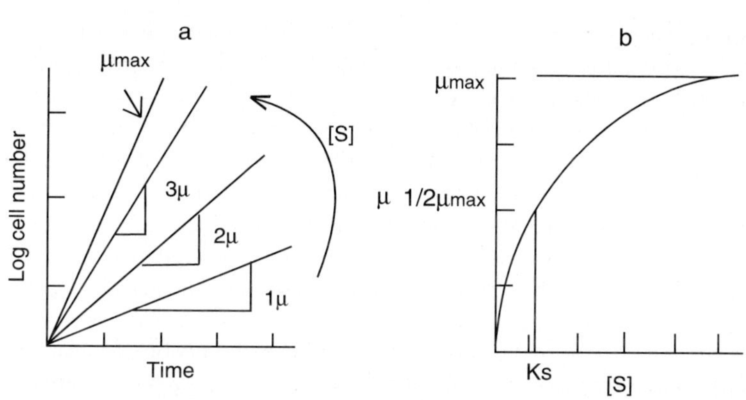

Fig. 2.1.2 Relationship between the concentration of the limiting nutrient and specific growth rate. *Panel* **a** indicates the growth as a function of time at different concentration of limiting nutrient. The relationship between the growth rate constant and the concentration of limiting nutrient is given in *panel* **b**

Table 2.1.1 Growth properties of anaerobic cultivation of *Saccharomyces kluyveri* and *S. cerevisiae* (data obtained with permission from Moller and Pisker 2001)

Growth parameter	*S. kluyveri*	*S. cerevisiae*
Growth rate (μ_{max}, [h^{-1}])	0.24	0.41
Biomass (Ysx [g/g])[a]	0.089	0.092
Ethanol (Yse [g/g])	0.350	0.376
Carbon dioxide (Ysc [g/g])	0.389	0.397
Glycerol (Ysgly [g/g])	0.109	0.107

[a] Yield coefficient (Y) is expressed as grams of biomass, ethanol, glycerol, carbon dioxide per gram of glucose consumed

phase, Xo is the dry weight of the inoculum and S is the concentration of limiting nutrient (in mg/L). Growth yield can also be expressed as the dry weight in grams of biomass formed per mole of substrate (Table 2.1.1). The parameters such as μ, Ys or Ks allow us to compare the growth performance of different strains or same strains under different growth conditions. For example, μ_{max} h^{-1} of *Candida tropicalis* and *Saccharomyces cerevisiae* is 0.74 and 0.47, while Ymolar (gm dry weight/mole glucose) of *Zymomonas mobilis* and *Saccharomyces cerevisiae* is 9 and 29, respectively. These comparisons provide insights into how organisms have optimized these parameters to remain competitive in a constantly changing environment.

2.1.4 Metabolic Strategy

As mentioned before, yeast prefers to obtain energy by fermenting glucose and not by oxidation, despite the fact that oxidation provides more energy per glucose as compared to fermentation. In fact, during fermentation on glucose, mitochondrial oxidation machinery is severely suppressed. Not only that, a significant fraction of cells in the population sporadically lose mitochondria during growth on glucose. After exhaustion of glucose, subsequent growth occurs due to mitochondrial oxidation of ethanol, which is accumulated during the fermentative stage. Cells that have spontaneously lost mitochondria during growth on glucose form small colonies on solid medium and are referred to as petites. The small colony size of petites is because of their inability to use ethanol after glucose is exhausted from the medium. This is referred to as "petite-positive phenotype" (Box 2.1.1). This phenotype is not exhibited when cells grow on other fermentative carbon sources such as galactose, nor is it exhibited by other species of *Saccharomyces*. For example, *Saccharomyces kluyveri*, a close relative of *Saccharomyces cerevisiae*, can grow anaerobically, but cannot survive if mitochondria are lost, and therefore is referred to as "petite negative yeast".

What is the teleological reason for the sporadic loss of mitochondria when yeast grows on glucose? *S. kluyveri*, and *S. cerevisiae*, when grown on glucose under anaerobic condition, display similar growth parameters except the μ_{max} (Table 2.1.1). This suggest that in *S. cerevisiae*, the metabolic energy derived from fermentation of glucose is diverted for cell multiplication than for maintaining mitochondria, at least in a fraction of cell population. This could account for the overall higher μ_{max} of *S cerevisiae* as compared to *S. kluyveri*. In evolutionary terms, it may mean that the petites, which have a disadvantage once glucose exhausts, seem to "sacrifice" their growth on ethanol for achieving higher growth rate for the common good.

Fermentation of glucose by yeast reveals a near-perfect metabolic design to remain competitive. First, because fermentation is an energy-inefficient process compared to oxidation, it has to consume more glucose per cell division. In this context, high growth rate would result in faster depletion of glucose from the

Box 2.1.1 *Petite-positive phenotype*

During growth on glucose as the carbon source, cells of *Saccharomyces cerevisiae* constantly produce mutants characterized by reduced colony size and referred to as petites. Petite mutants, a special class of respiratory-deficient mutants, either lack a part or whole mitochondrial genome. Yeast groups such as *S. cerevisiae* that give rise to petite mutants without any apparent selective pressure are said to exhibit petite-positive phenotype while those that cannot generate petites are said to exhibit a petite-negative phenotype.

Saccharomyces kluyveri can grow anaerobically but is not petite-positive. That is, it cannot lose mitochondria. On the other hand, *Kluyveromyces lactis*, a close relative of *Saccharomyces*, ferments glucose to ethanol, but neither exhibits petite-positive phenotype nor can grow anaerobically. In the latter two cases, mitochondrial function is absolutely essential. It is suggested that the petite-positive phenotype of *Saccharomyces cerevisiae* evolved due to the reorientation of metabolism as a consequence of genome duplication followed by rearrangement of genes.

Table 2.1.2 Doubling time of normal and petite strains of *Saccharomyces cerevisiae* on hexoses (data obtained with permission from Deken 1966)

	Wild-type		Pettite	
Carbon	Doubling time (in min)	Fermentation ($\mu l CO_2$/10 min/10^7 cells)	Doubling time (in min)	Fermentation (μl CO_2/10 min/10^7cells)
Glucose	53	78.0	70	72.4
fructose	53	69.0	70	70
Mannose	63	46.0	96	52.0
Galactose	72	15.3	139	30.0

medium, making it unavailable for the competing organism. It has been estimated that the rate of glucose uptake is 2×10^7 molecules per second per cell. Remember, that during fermentation, 1/3 carbon of glucose is preserved as ethanol for future use. Unlike yeast, ethanol is toxic to many microorganisms and cannot use ethanol as a carbon source. However, once glucose is exhausted, mitochondrial activity is derepressed, and yeast switches over to oxidative mode to consume ethanol as the source of energy and carbon. Thus, *Saccharomyces cerevisiae* owes its competitiveness to a combination of several features that evolved over millions of years.

Unlike growth on glucose, growth on galactose is an expensive affair as it necessitates synthesis of Leloir enzymes, which constitute approximately 5% of total cellular proteins when cells grow on galactose as the sole carbon source (Box 2.2.2). This overwhelming energy demand probably cannot be met by fermentation alone, which yields just two ATP/galactose consumed. This is consistent with the observation that mitochondria-less yeast grow at a rate half that of

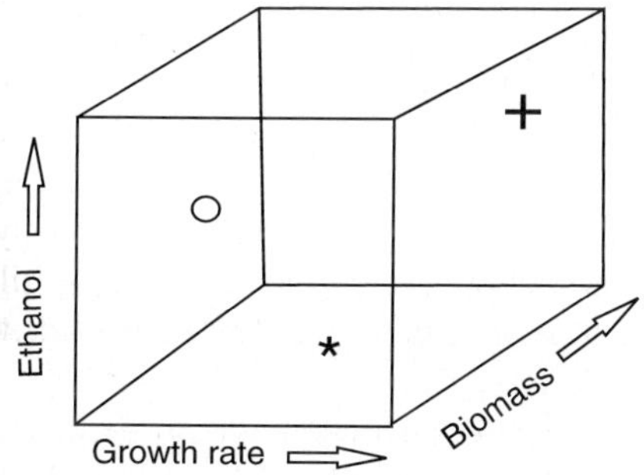

Fig. 2.1.3 Schematic illustration of metabolic space. Hypothetical metabolic space bounded by growth rate, ethanol production and biomass production. Organism represented by "O" has optimized high ethanol production and biomass production but low growth rate, "*" has optimized high growth rate but low biomass and ethanol production while "+" has optimized high ethanol and biomass and growth rate. Parameters are optimized as dictated by the evolutionary trajectory taken by the organisms. For example, humans can be considered to have optimized complex functions at the expense of growth rate

Box 2.1.2 *Fermentation*

Fermentation has been loosely used to indicate large-scale cultivation of microorganisms for industrial purposes. It has also been used as a synonym for respiration in the absence of oxygen. In fermentation, no external electron acceptor is required and redox reactions are balanced internally. Moreover, carbon is not completely oxidized. In fact, some industrial processes are aerobic and involve complete oxidation of the carbon source. Respiration in the absence of oxygen, that is, anaerobic respiration, differs from fermentation in that an external electron acceptor is used. For example, certain organisms use NO_3 as an electron acceptor and reduce it to NH_3 or SO_4 is reduced to H_2S.

Box 2.1.3 *Biosynthetic rate*

There is an inverse correlation between rates of metabolism and the size of organisms. If we take the rate of metabolism of humans as say 1, then elephant has 0.2, mouse has 10, and yeast has 100. This difference is essentially due to the large surface area/volume or surface area/weight. This enables the microorganisms to exchange matter and energy very efficiently. For example, a 200-lb. pound man has a surface area of $24{,}000\,cm^2/10{,}000\,g = 2.4\,cm^2/g$. A bacterium $1 \times 10^{-7}\,cm^2/2 \times 10^{-12}\,g = 50{,}000\,cm^2/g$.

The rate of protein synthesis is an index of biosynthetic activity. The protein biosynthetic rate in aged, adult, young adult, and infants is 1.9, 3.0, 6.9, and 17.4 g/kg/day, respectively. Assuming 50% dry weight is protein and 12% is the total dry weight, 60 g is protein/kg of cells. A bacterial cell doubles itself every 30 min.

That is 1 kg of bacterial cell becomes 2 kg and 4 kg. At the end of 1 h, 3 kg of biomass is produced, which is equivalent to 180 gm/kg/h.

the strains containing mitochondria (Table 2.1.2). It is intriguing to know that while glucose and galactose have the same free energy content, the design of their metabolism is vastly different.

Unlike yeast, humans depend on mitochondrial oxidation for energy demands. However, red blood cells (RBC) derive energy exclusively by fermentation of glucose to lactate. This adaptation ensures that RBC does not oxidize glucose using oxygen which is meant to be supplied to other tissues. This is an example of metabolic differentiation to ensure efficient transport of oxygen from lungs to tissues. Skeletal muscles also ferment glucose to lactate when mitochondrial oxidation is unable to keep pace with the influx of glucose under conditions of vigorous muscular activity. This is an example of physiological adaptation. What is the metabolic fate of lactate?

Lactate finds its way into the liver through blood circulation where it gets converted to glucose, which is released back into the blood circulation. This is similar to the metabolic strategy adapted by yeasts during fermentation.

References

Wilkinson JF (1986) Introduction to microbiology. In: Both IR, Gooday GW, Gow NAR, Hamilton WA, Prosser JI (Eds) Basic microbiology series, vol. 1. Blackwell Science Publications, Oxford

Moller K, Olsson L, Piskur J (2001) Ability for anerobic growth is not sufficient for development of the petite phenotype in *Saccharomyces kluyveri*. J Bacteriol 183:2484–2489

Deken RH (1966) The Crabtree effect: a regulatory system in yeast. J Gen Microbiol 44:149–156

Prosser JI (1995) Kinetics of filamentous growth and branching. In: Gow AR, Gadd GM (eds) The growing fungus. Chapman and Hall, London, pp 301–335

Scheeler P, Bianchi DE (1987) Cell and molecular biology. Third edition. John Wiley and Sons (Asia) Pte. Ltd.

Van Uden N (1971) Kinetics and energetics of yeast growth. In: Rose AH, Harrison JS (eds) The yeasts, vol. 2. Academic Press, New York, pp 75–118

2.2 Enzyme Adaptation

2.2.1 Introduction

Current understanding of the concept of differential regulation of gene expression, which is fundamental for the understanding of a whole range of biological processes such as development, differentiation emerged from a detailed analysis of the nature of enzyme adaptation. Enzyme adaptation was initially observed in microbial systems as a phenomenon of switching from one metabolic state to another in response to the presence of specific substrates. In this section, regulation of galactose metabolism in yeast is discussed in the context of brief historical perspective of enzyme adaptation. This gives a glimpse of intellectual and experimental efforts directed at understanding the phenomenon of enzyme adaptation. This paradigm continues to provide insights into the working of not just transcriptional regulation but also helps us understand many rapidly evolving concepts of modern biology.

2.2.2 Adaptation to Nutrients

As early as the 1890s, Frederic Dienert discovered that yeast pre-grown on glucose starts utilizing galactose with a delay, but yeast pre-grown on galactose starts using glucose or galactose without delay. Further, if yeast grows on a mixture of glucose and galactose, it first ferments glucose to ethanol and temporarily ceases growth before it starts fermenting galactose to ethanol. This effect was called "the glucose effect". During the course of these investigations, he also identified yeast strain unable to use galactose. By the turn of the 20th century, similar observations were rediscovered in bacteria. Henning Karstrome invoked the idea of enzyme adaptation to explain the delay in acclimatization when microbes start utilizing alternate carbon sources. He referred to enzymes existing in a living cell regardless of the nature of the nutrients present in the medium as "constitutive" while those formed only in the presence of their pathway substrate such as galactose as "adaptive".

In 1938, Yadkin proposed a conceptual basis for enzyme adaptation and suggested that enzymes exist in equilibrium between active and inactive form. The equilibrium is in favor of inactive form for adaptive enzymes while it is the opposite for the constitutive enzymes. He further suggested that when the adaptive enzyme which exists in inactive form comes in contact with the substrate, the equilibrium shifts towards the active form. This theory was referred to as the mass-action theory of enzyme adaptation. The view that the substrate somehow influences the protein to change its activity was also used to explain the diversity of antibodies. Sol Speigelman (who was to later spend considerable effort in understanding the "long-term adaptation" phenotype in yeast, see below) proposed the "plasmagene" hypothesis to explain adaptation. According to this hypothesis, the substrate would induce duplication of the relevant genes to increase the enzymes. This idea did not stand the test of scientific scrutiny and was quickly abandoned.

In the 1940s, Jacques Monod observed that in certain mixtures of carbon sources, *E. coli* showed single growth cycle while in others it showed two cycles of growth separated by temporary cessation of growth. He termed this phenomenon "diauxie" (Fig. 2.2.1). The adaptation of an enzyme system required to catabolise galactose occurred in the absence of cell division was first observed in yeast as early as 1900. Later, a similar observation was also made in *E. coli*. This indicated

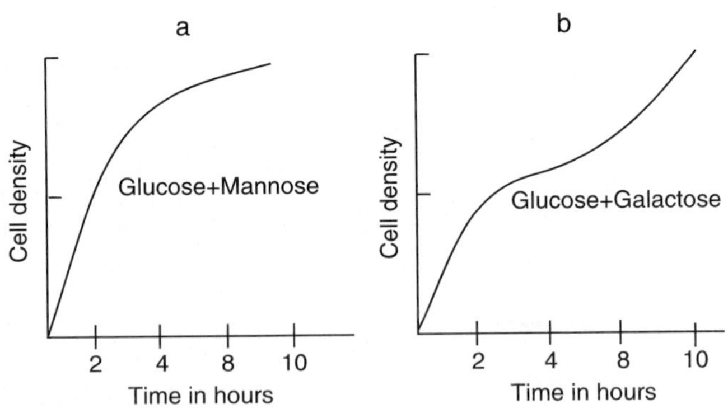

Fig. 2.2.1 Schematic illustration of growth profiles of *E. coli* in glucose medium either with mannose **a** or with galactose **b**. Note that in the presence of mannose and glucose there is only one exponential phase while in the presence of glucose and galactose the exponential phase is separated by a lag phase

that enzyme adaptation is not due to an alteration in the genetic structure, since the latter occurs only during cell multiplication. Second, the phenomenon of adaptation was sensitive to the presence of energy uncouplers, indicating that the expenditure of energy is a prerequisite for adaptation. These results convinced Monod that enzyme adaptation is due to the delay in the synthesis of enzymes rather than a delay in their transformation in the presence of the substrate. The challenge, however, was to relate the role of the substrate and the gene to account for the fresh synthesis of enzyme molecules. By 1960 Monod provided a molecular basis of enzyme induction, the cornerstone for our present understanding of regulation of gene expression. However, enzyme adaptation, which Dienert observed with respect to galactose utilization in yeast, took a curious turn.

2.2.3 Long-Term Adaptation

After the initial discovery of enzyme adaptation by Dienert, yeast played a key role as an experimental organism in the elucidation of glycolysis, but its use in genetic studies was a suspect for long time due to the non-Mendelian segregation pattern in genetic crosses. In the 1930s, Ojvind Winge began research in yeast genetics and showed that the yeast life cycle involves an alternation between haploid and diploid phase (discussed in the previous chapter). While investigating the ability of yeast strains to ferment sugars, Winge and Roberts encountered an unusual yeast, which took as many as 3–4 days to adapt to galactose as compared to few hours for a normal strain. This was referred to as "long-term adaptation" (Fig. 2.2.2).

An unusual feature of this phenotype was that galactose-adapted cells on subsequent exposure to galactose do not show long-term adaptation. However, if

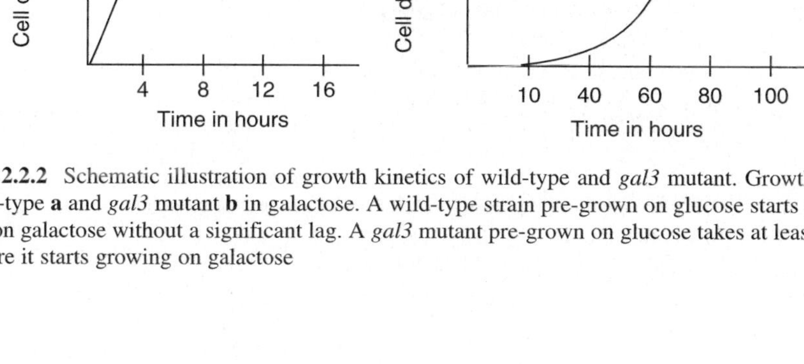

Fig. 2.2.2 Schematic illustration of growth kinetics of wild-type and *gal3* mutant. Growth of a wild-type **a** and *gal3* mutant **b** in galactose. A wild-type strain pre-grown on glucose starts growing on galactose without a significant lag. A *gal3* mutant pre-grown on glucose takes at least 48 h before it starts growing on galactose

galactose-adapted cells are cultivated in the absence of galactose for few generations, they lose the ability to rapidly adapt to the subsequent exposure to galactose. That is, during a few generations of growth on carbon sources other than galactose, the mutant strain loses the ability to rapidly adapt to galactose. Therefore, these cells are not only defective in responding quickly to galactose, but are unable to retain the property of rapid adaptation acquired during growth on galactose (Fig. 2.2.2). Preliminary genetic analysis indicated that it is a recessive defect at a genetic locus designated as *GAL3*. Following this discovery, Speigelman and co-workers conducted a detailed analysis of long-term adaptation.

2.2.4 *Single-Cell Analysis of Long-Term Adaptation*

This phenotype provided a convenient experimental system for analyzing the phenomenon of enzyme adaptation. As yeast divides by budding, it is possible to monitor whether the mother and the successive daughters (daughters produced form the same mother) retain rapid induction phenotype when exposed to glucose. For this purpose, a *gal3* cell adapted to galactose is maintained in a glucose medium and buds are removed as and when they are formed. The ability of these buds and the mother cell to respond to galactose is independently assessed by transferring to galactose medium (Fig. 2.2.3).

It appeared that factors acquired by *gal3* cells during adaptation to galactose were reduced during the subsequent growth in the absence of galactose. A positive

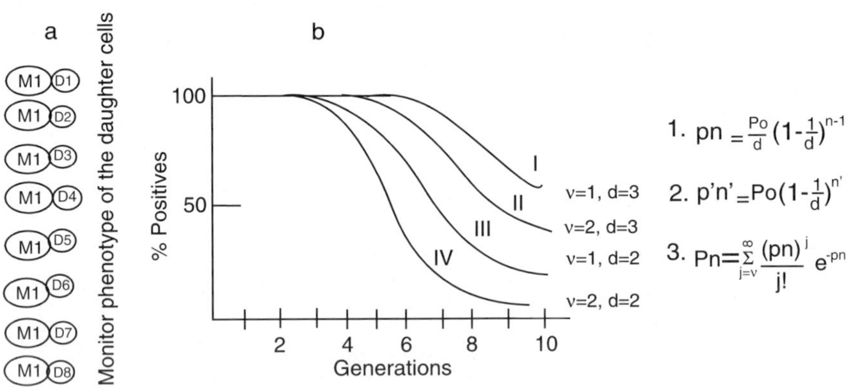

Fig. 2.2.3 Single-cell analysis of LTA. **a** The experimental strategy for determining the de-induction using single-cell analysis. **b** Theoretical curves generated based on statistical analysis. The average number of elements remaining in an *n*th generation daughter cell (Eq. 1), the average number of elements remaining in a mother cell after it has produced *n* daughter cell (Eq. 2), and the expected proportion of positive cells among the *n*th generation buds (Eq. 3) can be calculated. Po is the number of elements initially present, 1/d is the fraction of the parental elements that pass into the daughter cell and υ is the minimal number of elements required for the cell to be positive (adapted with permission from Speigelman et al. 1950)

mother cell itself becomes negative after six to seven generations. In some cases, the mother cell receives more while in other cases the daughter cell receives more. Overall, the data suggest that after a certain number of divisions have occurred, the number of elements available for distribution is such that the two cells produced as a result of division cannot both be positive. It is observed that the rapid induction and long-term adaptation phenotype of the daughter cells can alternate in successive generations. For example, in pedigree 2 (see pedigree 2, Table 2.2.1), the sixth and eighth buds show long-term adaptation while the seventh bud shows rapid induction. Table 2.2.2 gives the consolidated pedigree data.

This data is amenable for quantitative analysis and the proportions of the positives found in each generation among the mother and daughter cell can be determined. With a constant number of inducing elements present in a galactose adapted *gal3* cells, the proportion of positives to be expected at any given generation depends upon the following parameters. Po, the number of elements initially present; 1/d the fraction of elements that pass on to the daughter cell and ν the minimum number of elements required to yield the positive phenotype. From the equation shown in Fig. 2.2.3, theoretical curves can be generated for different proportions of the positives as a function of generation for different values of 1/d and ν. It turned out that the experimental data fits with the III curve. Of the three parameters, variation in Po would only alter the number of generations before the appearance of negative cells and this will not change the shape of the descending part of

Table 2.2.1 Single-cell analysis of four individual pedigrees (data obtained with permission from Speigelman et al. 1950)

Pedigree	Generations									Mother cell
	1	2	3	4	5	6	7	8	9	
1	+	+	0	+	+	+	–	–	0[b]	–[a]
2	+	+	+	+	+	–	+	–	–	–
3	+	+	+	+	+	+	–			+[a]
4	+	+	+	+	+	+	+			

[a] "+" and "–" signs indicate that the clone derived from the single cell exhibits rapid or slow induction, respectively

[b] "0" indicates that the clone did not survive and a blank space indicates that bud isolation was not continued

Table 2.2.2 Summary of the pedigree analysis (data obtained with permission from Speigelman et al. 1950)

Generations	Positives	Negatives	Total	Positives (%)
2	34	0	34	100
3	37	0	37	100
4	27	0	27	100
5	27	1	28	96.5
6	26	6	32	81.5
7	22	12	34	65.0
8	9	18	27	33.3
9	3	10	13	23.0

the curve. On the other hand, 1/d and ν influence the descending part of the curve in opposite direction. Therefore, the value of 1/d=2 and ν=1 is not the only value that would fit the experimental data, other combinations of 1/d and ν value would also fit the data. However, a more detailed analysis suggested that the value of ν=1 and 1/d=2 and Po was calculated to be 200.

Above analysis demonstrated that in *gal3* cells, galactose eventually induces a factor required for induction. If galactose is withdrawn, it gets diluted below a threshold required for rapid induction within six to seven generations. It was also observed that in glucose grown population of *gal3* mutants, one out of approximately 1,000 cells is a true galactose fermentor. Accordingly, it is the time required for the multiplication of this small fraction of cells that causes the delayed growth and not because of the slow induction of the factor. Based on these results, Speigelman invoked the concept of heterogeneity in cell population as a mechanism of adaptation to galactose. Recent analysis, in fact, supports the view that LTA is due to the cellular heterogeneity (see section 8.2.3).

2.2.5 Galactose Metabolism

The galactose metabolic pathway is commonly referred to as the Leloir pathway after Luis Federico Leloir, who discovered that the intracellular galactose is converted to glucose through four distinct enzymatic reactions. Galactokinase catalyses the conversion of intracellular α D-galactose to galactose-1-phosphate. Galactose-1-phosphate is converted to glucose-1-phosphate by uridyl transferase. UDPgluocse needed for this reaction is replenished by the conversion of UDPgalactose to UDP glucose by the epimerase. Glucose –1-phosphate is then converted to glucoe6-phosphate by phosphoglucomutase. As phosphoglucomutase is also involved in converting glucose-6-phosphate to glucose-1-phosphate during growth on glucose, it is not generally considered as a member of Leloir pathway. The need for three enzymes for epimerising galactose to glucose is quite unique in biochemistry.

Yeast utilizes melibiose, a disaccharide consisting of glucose and galactose linked through α-glycosidic linkage. α-galactosidase cleaves melibiose into glucose and galactose, which is taken up by yeast as carbon sources. *Saccharomyces cerevisiae* strains normally do not code for α-galactosidase, but *Saccharomyces cerevisiae* strains containing α-galactosidase have been derived by interspecies crossing with *Saccharomyces carlsbergensis*. Unlike the Leloir enzymes, α-galactosidase is an extracellular enzyme, and its expression is controlled by the same mechanisms as Leloir genes.

Free galactose exists as an equilibrium mixture of α and β forms and it is the α form that is the substrate for galactokinase. Aldose 1-epimerase (EC 5.1.1.3) or mutarotase interconverts these two forms. While in *E. coli* and humans, this enzyme is encoded by a distinct gene, in yeast, mutarotase is a part of the epimerase polypeptide.

The genetic basis of galactose metabolism was first demonstrated by Lindegren and Lindegren by conducting genetic analysis of haploid strains defective in galactose fermentation. It was believed that the sequential induction of enzyme activity

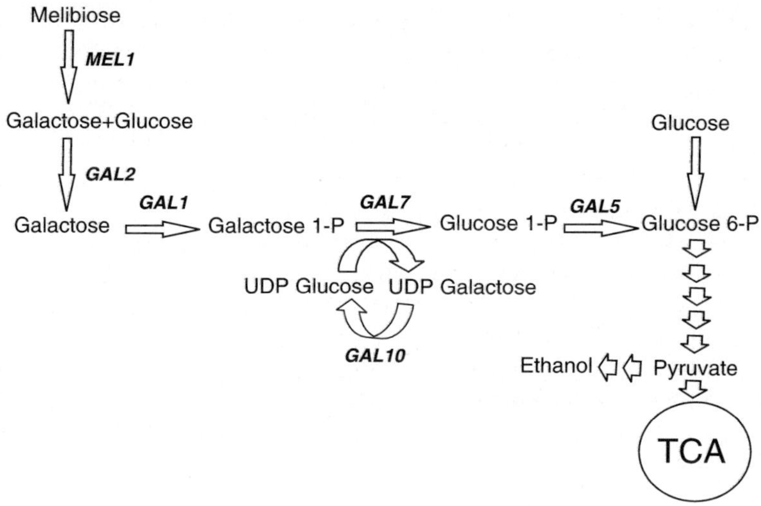

Fig. 2.2.4 Galactose metabolism in yeast. *GAL1*: Galactokinase (EC 2.7.1.6); *GAL7*: Galactose-1phosphate uridyl transferase (EC 2.7.7.12); *GAL10*: Uridine diphosphoglucose 4-epimerase (EC 5.1.3.2) *GAL5*: (*PGM2*) Phosphoglucomutase (EC 2.7.5.1) ; *MEL1*: α-galactosidase (EC 3.2.1.22,) *GAL2*: galactose permease

occurs in response to the formation of a product which in turn acts as an inducer for the subsequent enzyme. Contrary to this expectation, galactose induced the activity of uridyl transferase and epimerase in a galactokinase-less mutant yeast strain. This study indicated that free galactose induces not just galactokinase but also the activities of all the three Leloir enzymes. Leloir enzymes were purified from cell-free extracts obtained from galactose adapted cells. Antibodies raised against these proteins were used as probes to determine the mechanism of galactose activation, which is discussed in the next chapter.

Box 2.2.1 *Determination of enzyme activity*

Leloir enzymes were purified using conventional protein-purification techniques. Purification of enzymes from a complex mixture of proteins requires an assay method to monitor the presence of the enzyme in fractions obtained during the purification. As an example, different methods for detecting galactokinase activity are discussed.

1. *Colorimetric method.* This method takes advantage of the fact that free galactose concentration decreases as the reaction proceeds. Free galactose concentration present in the reaction mixture after a specified time point is monitored by allowing it to react with 3,5, dinitrosalicylic acid. This oxidizes the free reducing sugar (R-CHO) to the corresponding acid

(continued)

Box 2.2.1 (continued)

(R-COOH) and in the process is reduced to 3 amino, 5 nitrate salicylate. The concentration of this can be determined from the molar extinction coefficient by recording absorption at 575 nm.

2. *Coupled assay*. ADP formed during the reaction is coupled to the conversion of phosphoenolpyruvate to pyruvate in the presence of pyruvate kinase. The pyruvate formed is then coupled to the formation of lactate from pyruvate in the presence of lactate dehydrogenase. NADH oxidation due the conversion of pyruvate to lactate is monitored by recording a decrease in absorbance at 340 nm. The decrease is proportional to pyruvate formed, which in turn is proportional to the ADP formed in the galactokinase reaction.
3. *Radioactive assay*. This assay takes advantage of the fact that galactose-1-phosphate formed can be separated from free galactose by adsorbing to a charged surface such as DEAE filter paper. For this purpose, ^{14}C labeled galactose, instead of normal galactose is used. The ^{14}Cgalactose1-phosphate present in the reaction is separated by loading the reaction mixture onto DEAE paper strips followed by washing with excess water. During this step, radioactive galactose-1-phopshate retained on to the paper as it is charged while uncharged galactose is washed off. The filter paper is counted for the radioactivity. Other charged molecules such as ADP and unreacted ATP would also be retained as they are charged, but this will not interfere as the presence of only radioactivity is monitored.

Box 2.2.2 *Energetics of galactokinase synthesis*

A yeast cell yields ~6×10^{-9}mg of protein. One milligram of total protein extracted from galactose-adapted yeast contains sufficient galactokinase to catalyze 45 µM of galactose to galactose-1-phosphate in an hour. The turnover number of yeast galactokinase is 57/s per enzyme molecule. Based on this, the concentration of galactokinase in a yeast cell is determined to be in the nanomolar range. Its amino-acid sequence has also been deduced from its gene sequence. One molecule of galactokinase has 1,815 carbon atoms. If yeast grows on galactose as the sole source of carbon, this is equivalent to 302 galactose molecules. Consider the use of galactose as a source of carbon and energy to make a molecule of galactokinase. A total of 302 equivalent of galactose molecules are required to supply just the carbon alone. Assuming that four ATPs are required for one peptide bond formation, 2,108 ATPs are required for the synthesis of one molecule of galactokinase starting from amino acids. Here, the number of ATP required for the synthesis of amino acids is not considered. These calculations give us a glimpse of the energetics of galactose utilization. Remember, yeast also has to divert the carbon and the energy derived form galactose to other cellular activities when it grows on galactose as the sole source of carbon.

Box 2.2.3 *Galactose metabolic pathway is evolutionarily conserved*

In humans, the Leloir pathway of galactose metabolism is especially important during early childhood since galactose is one of the major sources of energy. In milk, galactose exists as a component of disaccharide lactose. This is absorbed as glucose and galactose after its hydrolysis by β-galactosidase, an enzyme present in the intestine. An individual bearing defect in galactokinase suffers from juvenile cataracts due to the accumulation of galactitol derived from un-metabolized galactose. Withdrawal of galactose from the diet of such individuals alleviates the symptoms considerably. Lack of transferase shows severe physiological disturbance due to the accumulation of galactose-1-phossphate. This leads to physiological disturbance such as ovarian dysfunction, learning disabilities, and liver enlargement. Due to the endogenous synthesis of galactose, this defect cannot be alleviated even upon withdrawing galactose from the diet. Individuals bearing the above defects occur in the population at a frequency of 1 in 30,000. Individuals lacking epimerase are very rare, indicating that its function might be essential.

References

De Robichion-Szulmajster H (1958) Induction of enzymes of galactose pathway in mutants of *Saccharomyces cerevisiae*. Science 127:28–29

Frey PA (1996) The Leloir pathway: a mechanistic imperative for three enzymes to change the stereochemical configuration of a single carbon in galactose. FASEB J 462:461–470

Holden HM, Ratment I, Thoden JB (2003) Structure and function of enzymes of the Leloir pathway of galactose metabolism J Biol Chem 278:43885–43888

Holton JB, Walter JH, Tyfield LA (2000) In: Scriver CR, Beaudet Al, Sly SW, Valle D (eds) Metabolic and molecular basis of inherited diseases, 8th edn. McGraw Hill, New York, pp 1553–1587

Johnston M (1987) A model fungal gene regulatory mechanism: The GAL genes of *Saccharomyces cerevisiae*. Microbiol Reviews pp 458–476

Lindegren CC, Lindegren G (1947) Mendelian inheritance of genes affecting vitamin synthesizing ability in *Saccharomyces*. Ann Missouri, Botan Garden 34:95–99

Monod J (2003) From enzymatic adaptation to allosteric transitions. In: Ullmann U (ed) Origins of molecular biology. ASM Press, Am Soc Microbiol, Washington, DC, pp 295–317

Mortimer RK (1993) Ojvind Winge: founder of yeast genetics. In: Hall MN, Linder P (eds) The early days of yeast genetics. Cold Spring Laboratory Press, pp 3–16

Mujumdar S, Ghatak J, Mukherji S, Bhattacharji H, Bhaduri A (2004) UDP galactose 4 – Epimerase from *Saccharomyces cerevisiae*. A bifunctional enzyme with aldose 1-epimerase activity. Eur J Biochem 271:753–759

Muller-Hill B (1996) The lac operon: a short history of genetic paradigm. Walter de Gruyter, Berlin

Segal S (1998) Galactosemia today: the enigma and the challenge. J Inher Metab Dis 21:455–471

Sherman JR and Adler J (1963) Galactokinase from *E. coli*. J. Biol. Chem. 238:873–878

Spiegelman S, DeLorenzo WF, Campbell AM (1950) A single-cell analysis of the transmission of enzyme-forming capacity in yeast. J Bacteriol 37:513–523

Speigelman S, Sussman RR, Pinska E (1950) On the cytoplasmic nature of "long-term adaptation in yeast". Proc Natl Acad Sci USA 36:591–605

Timson DJ, Reece RJ (2003) Sugar recognition by human galactokinase. BMC Biochem 4:1–8

Winge O, Roberts C (1948) Inheritance of enzymatic characters in yeasts and the phenomenon of long-term adaptation. C.R. Trav. Lab. Carlberg. Ser Physiol 24:264–315

Yang J (2003) Studies in the substrate specificity of *Escherichia coli* galactokinase. Organ Lett 5:2223–2226

2.3 Induction of Leloir Enzymes

2.3.1 Introduction

We learned that Leloir enzyme activities are present only in yeast cells adapted to galactose but not in other carbon sources. The mechanism of how galactose increases the activities of these enzymes was not understood. The increase in enzyme activity could be due to either the activation of pre-existing enzyme by mechanisms such as posttranslational modification or to an increase in the absolute number of enzyme molecules per cell. An increase in the number of enzyme molecules could be a consequence of many factors, such as increased rate of transcription followed by translation, increase in mRNA stability, decreased rate of degradation of mRNA or protein or a combination of the above possibilities (Fig. 2.3.1, Box 2.3.1). In this section, experiments that demonstrated that galactose activates the synthesis of Leloir enzymes by increasing the steady-state concentration of transcripts of the corresponding genes are discussed.

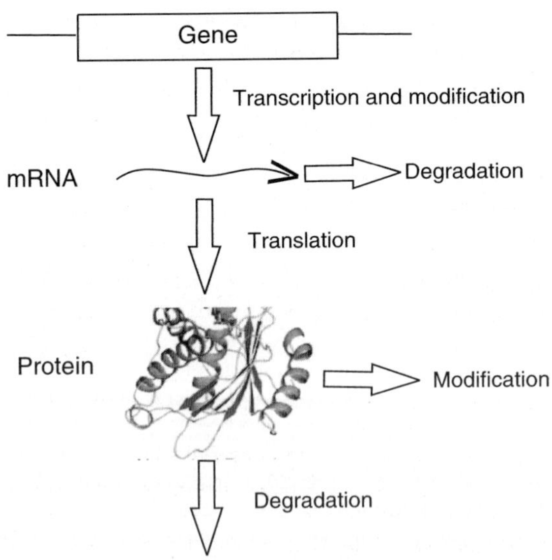

Fig. 2.3.1 Schematic representation of the regulation of gene expression. Any one or all of the above steps are potential targets for regulating the gene expression

Box 2.3.1 *Post-transcriptional and translational modifications*

Unlike prokaryotes, in eucaryotes, the primary product of mRNA, referred to as hnRNA, is processed to a mature functional form. hnRNA contains stretches of sequences called "introns", which need to be removed to give rise to functional mRNA products. In yeast, only a few protein-coding genes contain introns, while in humans, almost every protein coding gene has an intron. The number of introns vary from gene to gene. The introns are removed by a complex enzymatic process called "splicing", which occurs within the nucleus. Alternative splicing of the same hnRNA can give rise to different mature mRNA there by increasing the variation in the protein products formed per genetic unit. Another post-transcriptional modification of mRNA is processing of 5' and 3' ends. Generally, mRNA are capped at the 5′ end by 7 methyl guanosine and polyadynalated at the 3′ end. The above modifications are collectively called "posttranscriptional modifications". Similarly, the protein product can also undergo many chemical modifications. For example, phosphorylation, ADP ribosylation, methylation, proteolysis, ubuquitinylation, etc. Almost all of these processes are regulated, thus increasing the stringency of biological regulation.

2.3.2 Galactose Induces the Synthesis of Leloir Enzymes

Hopper and his group investigated whether galactose induces the synthesis of Leloir enzymes by monitoring the incorporation of radiolabeled leucine into uridyl transferase. Yeast cells were grown in galactose (inducing carbon source) or acetate (non-inducing and non-repressing carbon source) as carbon source in presence of radioactively labeled leucine as a tracer. Cell-free extract prepared from yeast cells was immunoprecipitated using antibodies raised against pure uridyl transferase and separated on sodium do-decyl sulphate polyacrylamide gel electrophoresis (SDS PAGE, see Box 2.3.2) followed by autoradiography (Fig. 2.3.2). They observed a band corresponding to the expected molecular mass of uridyl transferase only from extracts made from galactose but not acetate grown cells (Fig. 2.3.2) suggesting that galactose induces synthesis of uridyl transferase protein.

2.3.3 Galactose Activates the Transcription of GAL Genes

The increase in uridyl transferase synthesis in response to galactose could be due to a combination of reasons, such as increased transcription followed by translation or an increase in translation of the pre-existing mRNA. To distinguish between these possibilities, mRNA levels encoding uridyl transferase and galactokinase were monitored from a total mRNA population isolated from cells grown in acetate

Box 2.3.2 *In vitro translation*

This technique was developed to detect the presence of a specific mRNA. Wheat germ cell extract contains components necessary for translation of mRNA obtained from different sources. Wheat germ cell extract is processed to remove the endogenous amino acids and mRNA and supplemented with energy generating system required for protein synthesis. If exogenous mRNA is added to this lysate in the presence of labeled amino acids such as methionine or leucine supplemented with other cold amino acids, then synthesis of radioactively labeled proteins occur.

Radioactively labeled protein synthesis directed by the exogenously added mRNA present in the reaction mixture can be separated and analyzed by electrophoresis followed by detection. Before the advent of Northern blot analysis, this was the only technique available to detect a particular species of mRNA present in a heterogeneous population. It is now possible to monitor the expression of all the mRNAs simultaneously using microarray techniques, which will be discussed later.

Electrophoresis

This technique is one of the most widely used techniques for separating charged molecules such as proteins or nucleic acids. A sample containing a mixture of proteins or nucleic acids is separated through polyacrylamide or agarose gel under the influence of an electric field. Each molecular species migrates to a different extent depending upon its physicochemical properties. For example, nucleic acid is uniformly charged. That is, the charge-by-mass ratio remains the same regardless of the size, and therefore the separation is a function of the molecular weight. In proteins, it is possible that under native conditions, two proteins differing in molecular weight may have the same charge-by-mass ratio and would migrate to the same extent and separation would not occur. On the other hand, two proteins having the same molecular weight can differ in charge-by-mass ratio and could move to a different extent. By treating with sodium dodecyl sulphate (SDS), a denaturing agent, proteins can be conferred uniform charge-by-mass ratio. Therefore, SDS-treated proteins migrate through electrophoresis based on their molecular weight. Depending upon the experimental need, proteins can be separated either under native or under denaturing conditions. After separation, the samples are detected either by autoradiography or by staining, or both.

Staining and Autoradiography

After separation through electrophoresis, the gel is treated with a dye that imparts a specific color by interacting with nucleic acid or protein, which can be seen with the naked eye. For example, after separation through electrophoresis

(continued)

Box 2.3.2 (continued)

proteins are detected by staining with Coomassie blue, while nucleic acids are stained using ethidium bromides. For example, ethidium bromide intercalates with nucleic acid, which upon illumination at 280 nm, imparts characteristic fluorescence. The presence of radioactively labeled molecules can be detected by exposure to X-ray sensitive film. Here, the sample is kept in close contact with the film. An image corresponding to the position of the radioactive bands is imprinted on the film, which is detected after developing and fixing.

and galactose. The technique of detecting specific mRNA species in a mixture of heterogeneous population relies on the ability of wheat germ cell extract to support the translation of exogenous mRNA into proteins, in this case mRNA isolated from yeast cells. Total mRNA isolated from yeast cells grown in galactose and glucose were separately translated in wheat germ cell-free translation system in the presence of amino acids of which methionine is radioactively labeled. Immunoprecipitate obtained after treating the above reaction mixture with antibodies raised against galactokinase and uridyl transferase was subjected to SDS electrophoresis followed by autoradiography. Total mRNA isolated from galactose but not acetate grown cells directed the synthesis of a radiolabeled protein corresponding to the molecular weight of uridyl transferase (Fig. 2.3.3a, lanes 1 and 2) and galactokinase (Fig. 2.3.3b, lanes 1 and 2). This result indicated that mRNA directing the synthesis of galactokinase and uridyl transferase were present only in cells grown in galactose but not acetate.

To determine whether the same mRNA molecule encodes both galactokinase and transferase, total mRNAs isolated from galactose-grown cells were fractionated by sucrose density gradient centrifugation. During this procedure, mRNA gets

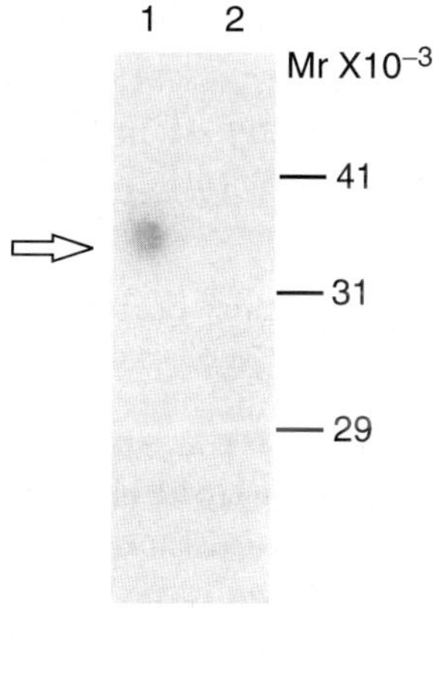

Fig. 2.3.2 Induction of synthesis of radioactively labeled uridyl transferase by galactose. Cell extract obtained from cells grown in presence of galactose (*1*) or acetate (*2*) with radioactively labeled leucine was immunoprecipitated by antibodies raised against uridyl transferase. Immunoprecipitate was separated on SDSPAGE followed by autoradiography (reproduced with permission from Hopper et al. 1978). *Arrow* indicates radioactively labeled galactose-1phosphate uridyl transferase

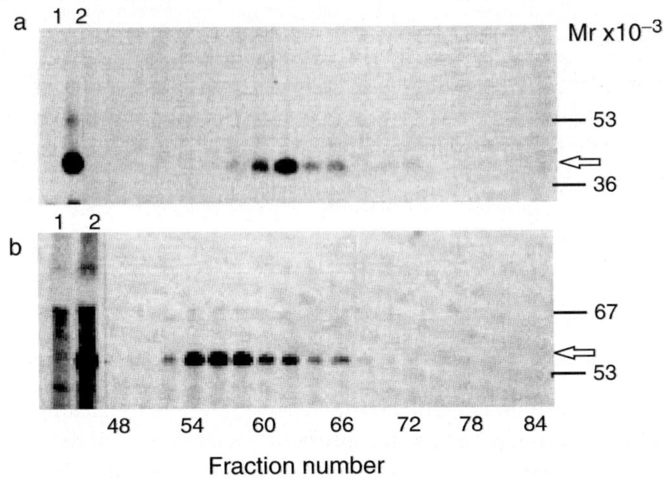

Fig. 2.3.3 Autoradiographic analysis of immunoprecipitated radiolabeled proteins obtained after in vitro translation. Total mRNA isolated from yeast cells grown in acetate (*lane 1*) and galactose (*lane 2*) were translated in vitro, immunoprecipitated by antibodies against uridyl transferase **a** or galactokinase **b**, and separated on SDS PAGE followed by autoradiography. In vitro translated products of mRNA obtained from alternate fractions collected after density gradient centrifugation (as indicated in the figure) are immunoprecipitated by antibodies against uridyl transferase **a** or galactokinase **b** and analyzed as before (adapted with permission from Hopper and Rowe 1978). Arrow indicates uridyl transferase **a** and galactokinase **b**

sedimented along the gradient depending upon the molecular weight. Fractions obtained after density gradient centrifugation were separately in vitro translated and immunoprecipitated and analyzed as before. Galactokinase and uridyl transferase synthesis were directed by mRNA present in different fractions obtained from density gradient centrifugation. If a single polycistronic mRNA were to code for both galactokinase and uridyl transferase, then they would have been translated from mRNA obtained from the same fraction, which was not what is observed. This indicated that mRNAs for galactokinase and uridyl transferase were transcribed from separate genetic units.

2.3.4 Galactose Activates a Genetic Program

Above results demonstrated that galactose activates the transcription of Leloir genes. We know that galactose is unable to do so if glucose is also included in the medium. How does galactose activate the transcription of these genes in the absence of glucose? How does glucose prevent galactose from activating the transcription of Leloir genes? The phenomenon wherein gene transcription is turned ON or OFF depending upon a specific stimulus is referred to as the "regulation of gene expression", which is one of the predominant mechanisms by which cells regulate their genetic potential for specific biological purposes. For example, a fertilized egg endowed with the complete repertoire of genetic program responds to different intra- and/or extracellular cues in a temporally and spatially regulated manner. A dried seed in response to water initiates a developmental program, which eventually leads to the formation of a full-fledged plant. Deciphering the molecular mechanisms of turning "ON" and "OFF" of genetic program starting from the primary step all the way through the manifestation of the phenotype is crucial for understanding the life process.

References

Hopper JE, Rowe LB (1978) Molecular expression and regulation of the galactose pathway genes in *Saccharomyces cerevisiae*. J Biol Chem 253:7566–7569

Hopper JE, Broach JR, Rowe LB (1978) Regulation of the galactose pathway in *Saccharomyces cerevisiae*: induction of uridyl transferase mRNA and dependency on GAL4 gene function. Proc Natl Acad Sci USA 75:2878–2882

Voet D, Voet JG (1990) Biochemistry. John Wiley and Sons Inc.

Wilson K, Walker J (2000) Practical Biochemistry Cambridge University press

Chapter 3
Genetic Dissection of Galactose Metabolism

3.1 Genetic Analysis of *GAL* Regulon

3.1.1 Introduction

How does yeast translate the three-dimensional information of galactose into a specific biological signal that activates transcription of *GAL* genes? Genetic analysis is the method of choice for exploring the molecular basis of such phenomena that are otherwise not amenable for routine biochemical analysis. The success of genetic analysis depends on the availability of a large number of genetic variants which have a discernible phenotype. In our example, yeast strains unable to utilize galactose as the sole carbon source, serve as a starting point for genetic analysis. This chapter describes the genetic analysis that led to the identification of genes responsible for galactose utilization.

3.1.2 Mutant Hunt

Genetic studies of galactose utilization were initiated by Carl Lindegren who isolated two haploid yeast strains that did not grow on galactose as the sole carbon source. Later, Douglas and coworkers showed that one of the above mutants is defective in galactokinase while the other was defective in galactose uptake. Further, they also conducted a systematic genetic analysis of galactose utilization by isolating large number of mutants defective in galactose utilization.

Frequency of occurrence of random genetic variants in a population of cells is too low, of the order of 10^{-6}, to be able to isolate the large number of mutants defective in galactose utilization, which is a pre-requisite for a comprehensive genetic analysis. This problem was circumvented by increasing the mutation frequency by treating yeast cells with mutagens such as ethyl methyl sulphonate. This protocol introduces mutations randomly and not necessarily only in genes involved in galactose metabolism. Cells that do not suffer from a defect in essential genes required for normal growth are recovered by allowing the mutagenized cell population to grow on permissive medium, for example, a complete medium containing glucose as the carbon source (Fig. 3.1.1). From this mutagenized pool, cells defective for galactose utilization were identified by replicating the cells onto plates containing only galactose as the carbon source (Fig. 3.1.1). Mutants unable to grow on galactose were present at a frequency of 10^{-3} and these were recovered from the master plate for further analysis. This approach of identifying and isolating mutant strains is referred to as a "genetic screen". One needs to isolate as many mutants as possible to ensure that the mutant population represents defects in all possible genes required for galactose utilization. These mutants are classified based on different criteria as is discussed below.

3.1.3 Segregation Analysis

The inability of a mutant to grow on galactose could be due to single or multiple gene defects. For example, a mutant strain defective in galactokinase as well as epimerase would phenotypically be no different than either a strain defective only in galactokinase or epimerase. However, they differ from one another in their genetic constitution. Mutant strains carrying more than one defective gene for galactose utilization interfere in genetic analysis and are weeded out by conducting segregation analysis, which is carried out by obtaining diploids by crossing wild-type and the mutant strains (Fig. 3.1.1b). If the diploid is capable of growing on galactose, the mutation is said to be "recessive", and if not, it is "dominant". While both recessive and dominant mutations are valuable for genetic analysis (see later chapters), we shall restrict only with recessive mutants. Diploids are sporulated and the individual spores of an ascus are separated with the help of the micromanipulator and tested for their ability to grow on galactose (see Fig. 3.1.2 for details). If asci show only $2^+:2^-$ spore pattern for galactose "growth: no growth" phenotype, then the mutant that formed the diploid has only one gene defect with respect to galactose growth (Fig. 3.1.1). If the mutant haploid harbors more than one defective gene for galactose growth, then the spores can receive any one of the defective genes, giving rise to segregation ratios such as $3^-:1^+$ or $0^+:4^-$ (Fig. 3.1.2) in addition to $2^+:2^-$. Only mutants that show $2^+:2^-$ segregation in the above analysis are considered for further studies.

3.1.4 Complementation Analysis

As mentioned before, mutations in any one of the genes required for galactose utilization leads to galactose-negative phenotype. Whether the galactose growth defect in any two independent mutant strains is the same or different is determined by complementation analysis. Haploid mutant strains of opposite mating type are crossed in pair-wise combination and the phenotype of the resulting dip-

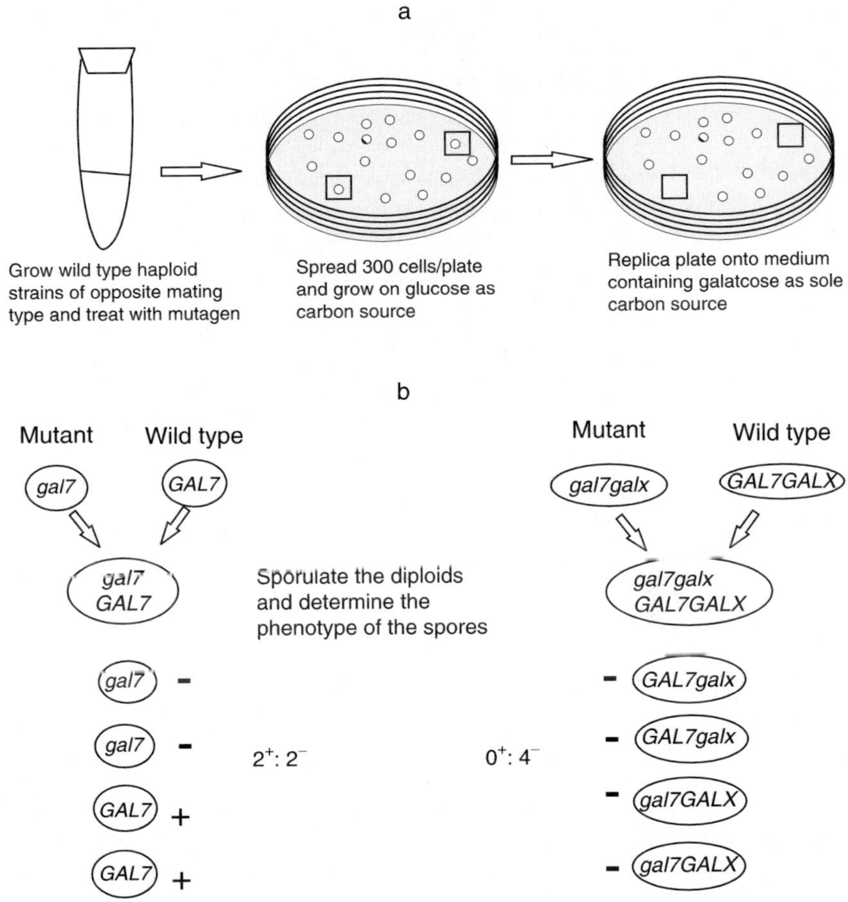

Fig. 3.1.1 Isolation of haploid mutants of both mating type and segregation analysis. **a** Scheme for isolating mutants defective in galactose utilization. Haploid strains of opposite mating type are separately treated with mutagen and screened for cells that have suffered mutation in galactose utilization pathway. The colonies represented by a *square* are a population of mutant cells that do not grow on galactose plates. The mutant cells from these colonies (cells obtained form a colony are genetically identical) are recovered and subjected to segregation analysis. **b** Diploids obtained by crossing individual haploid mutants (obtained from different colonies) and wild-type are sporulated and the phenotypes of the sister spores determined to identify the segregation pattern. + and – indicates growth or no growth on galactose, respectively. The *left panel* shows an ascus with $2^+:2^-$ segregation and the *right panel* shows the result of spore pattern obtained from a diploid formed by crossing a wild-type and a mutant haploid bearing two defective genes for galactose utilization (represented by *GAL7* and a hypothetical gene X). For the sake of clarity, only the $0^+:4^-$ pattern is shown, although other patterns such as $3^-:1^+$ can also be obtained

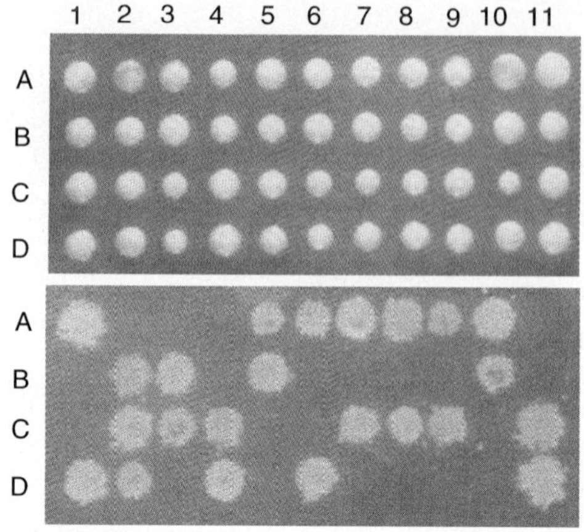

Fig. 3.1.2 Segregation analysis by tetrad dissection. The asci are streaked in the middle of a Petri plate containing nutrient agar. Spores from each ascus are physically separated and placed equidistant using a microscope fitted with a tetrad dissection micromanipulator. The sister spores (A, B, C, and D) obtained from different asci (1, 2, 3) are arranged as shown and allowed to develop into a colony by incubating at 30 °C for 2–3 days (*upper panel*) in rich agar medium. The colonies are then replicated on to the diagnostic medium to reveal the phenotype of the individual spores (*lower panel*). The asci dissected from a diploid with a heterozygous locus show 2^+2^- segregation except the ascus number 2, which shows aberrant segregation (3^+:1^-; for details see section 3.1.7) (adapted with permission from Sherman and Hicks 1991)

loid is determined. (1, 2, 3 and a, b, c, respectively, see Fig. 3.1.3) to grow on galactose was monitored (Table 3.1.1). If the diploid grows on galactose as the sole carbon source, then the haploids that constituted the diploid have defects in different genes required for galactose utilization. That is, two haploid mutants that have defects in separate gene complement while mutants that have defect in the same gene do not complement. Mutant strains that do not complement fall into the same complementation group (Table 3.1.2). In addition to the above, *GAL1*, *GAL2* (identified initially by Carl Lindgren) and *GAL3* (mutant isolated by Winge and Roberts) also represent separate complementation groups. Complementation analysis is possible only with recessive mutants. Dominant mutations do not complement because homozygotes and heterozygotes have the same phenotype.

Cell-free extract obtained from each mutant haploid strain induced with galactose (these mutants do not grow on galactose but are grown in ethanol and induced with galactose) were analyzed for Leloir enzyme activities. For example, all the members belonging to complementation group II did not express uridyl transferase, indicating that all the mutants belonging to this class have defective *GAL7*. All in all, mutants were categorized into seven different complementation groups: *GAL1, GAL2, GAL3, GAL4*, GAL5, *GAL7* and *GAL10*.

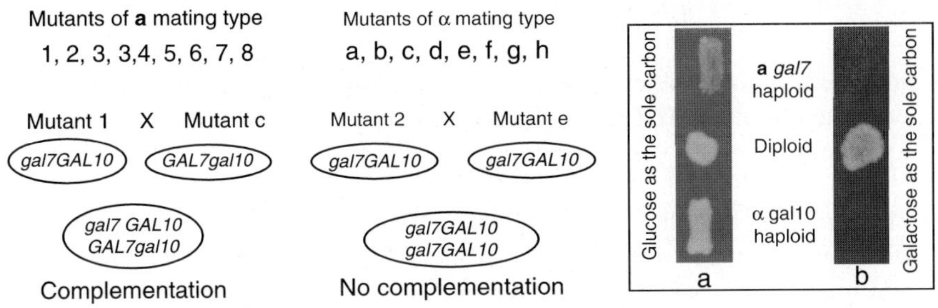

Fig. 3.1.3 Schematic illustration of complementation. Independent haploid mutants of opposite mating type bearing single gene defects are mated and the phenotype of the diploid is monitored. *Inset* shows the results of complementation between haploid **a** *gal7* and α *gal10* mutant. These mutants are separately streaked as well as patched together at the center of a plate **a** containing glucose as the sole carbon source. After 2 days of incubation, they are replicated on to a plate containing galactose as the sole carbon source. The patch of cells shown growing only at the center of the plate represent diploids growing on galactose due to complementation **b**. In the absence of complementation, the diploids would be present on the plate containing glucose but would not grow on galactose as the sole carbon source

Table 3.1.1 Complementation matrix

	a, d, e, f	b, c	g	h
1,5,6,7,8	–[a]	+	+	+
2,3	+[a]	–	+	+
4	+	+	–	+
9	+	+	+	–

[a]+ and – indicates complementation (diploids grow on galactose) and no complementation (diploids do not grow on galactose), respectively

Table 3.1.2 Complementation groups

Name of the mutants	Complementation group	Defective enzymes[a]	Locus
1, 5, 6, 7, 8, a, d, e, f	I	Mutase	*GAL5*
2, 3, b, c	II	Transferase	*GAL7*
4, g	III	All enzymes	*GAL4*
9, h	IV	Epimerase	*GAL10*

[a]Activity of Leloir enzymes was monitored in extracts obtained from each mutant grown in ethanol and induced by galactose

3.1.5 Concept of an Allele

Consider the four independent mutant strains belonging to complementation group II, represented by the *GAL7* locus. These mutant strains need not necessarily bear the same defects at the *GAL7* locus and therefore could represent the same or different alleles (although all of them have the same phenotype). If these strains bear different defects at the *GAL7* locus, then it means that five alleles of *GAL7* are identified. Why five and not four? A wild-type *GAL7* is also an allele. In general, different forms of the same gene are referred to as alleles. The word allele is used synonymously with gene. Even if thousands of alleles of a given locus exist in the population, a given haploid will have any one of the alleles, while a diploid can have any two.

How do we determine whether strains belonging to a complementation group carry the same or different alleles? For example, Oshima and his coworkers isolated 28 independent galactose-negative mutant strains which did not complement the previously isolated *gal4* mutants indicating that all the mutant strains are defective in *GAL4* (Fig. 3.1.4). It is possible that these mutant strains might bear different or the same alleles of *GAL4* locus. Comparison of the DNA sequence of these alleles provides direct evidence whether two mutant alleles are the same or not, but the concept of an allele had evolved even before the advent of DNA sequencing technology. Whether or not two strains bear the same or different alleles is determined based on recombination analysis. This is based on the phenomenon of genetic exchange that occurs between homologous chromosomes during meiosis by a process called homologous recombination (Fig. 3.1.4).

Here, recombinants are the haploid meiotic products whose genetic constitution is different than the haploids that generated the heterozygous diploid. If recombination occurs during meiosis as depicted (Fig. 3.1.4), then, of the two recombinant haploids, one would grow on galactose but not the reciprocal product. Therefore, the total number of recombinants would be twice the number of spores that grow on galactose. Frequency of recombination is a function of the distance between the mutant sites within different alleles.

It is obvious that no recombinants will be produced from haploids bearing the same alleles. A recombination frequency of 1% is equivalent to one genetic map unit. This is the same as 1 cM, the distance between any two mutations that yields on average 1% recombinant chromosomes or gametes or spore or sex cells. Using this approach, genetic length of the *GAL4* locus was determined to be on the order of 0.44% (Fig. 3.1.4). If we do not get even a single recombinant say in 1,000 meiotic products, we can only infer that most likely the mutations are separated by a distance less than 0.01 cM. That is, to infer that any two alleles are the same, one needs to screen a large number of meiotic products. Of the many *gal4* mutants, *gal4.62* was a nonsense mutant allele, since it was suppressed by a tRNA ochre suppressor. This was mapped to the middle of the *GAL4* locus and this mutant proved to be very useful in the genetic analysis of *GAL* system, which will be discussed later. The principle of recombination analysis discussed above is an extension of the genetic mapping tech-

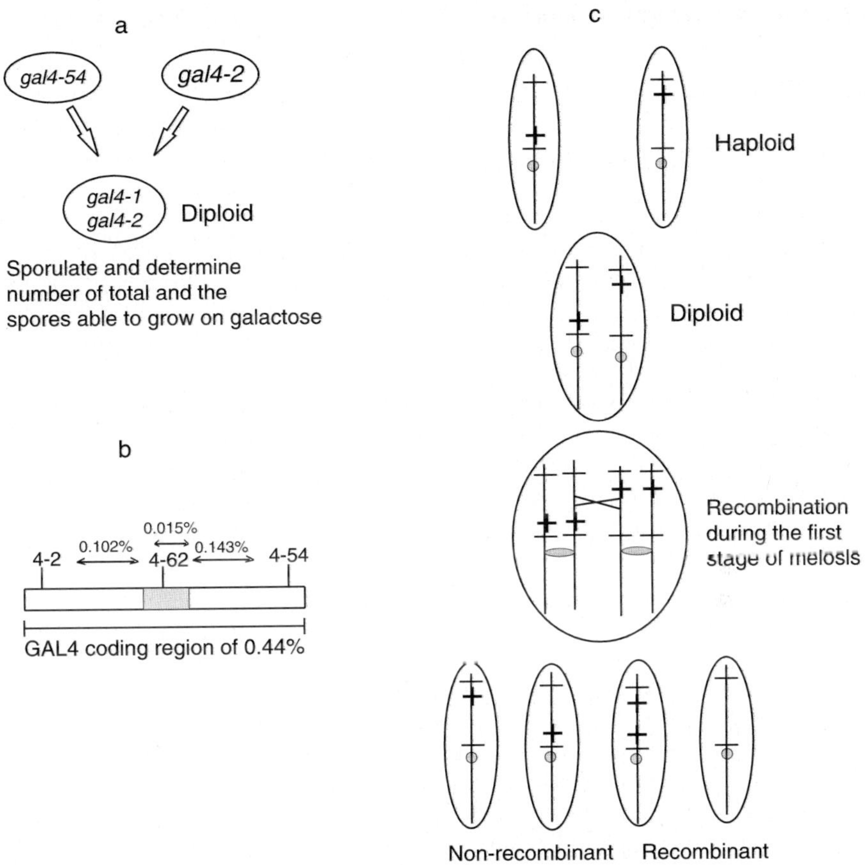

Fig. 3.1.4 Fine structure analysis of *GAL4*. **a** 23 recessive *gal4* mutants were crossed in all pairwise combination and the spores obtained from each group of diploids are separately pooled and subjected to random spore analysis as follows. The total number of spores in each pool are determined by plating the haploid spores on nonselective medium and GAL$^+$ spores are determined by spreading spores on a medium containing galactose as the sole carbon source. The number of spores able to grow on galactose multiplied by two gives the total number of recombinants (see text for details) and expressed as the percentage of the total. **b** This analysis allowed Matsumuto and co-workers (1980) to position the 23 independent mutant sites within *GAL4* gene. The position of only three mutant sites are indicated for the sake of clarity. The *numbers in percentages* indicate the recombinants obtained from a cross between the mutants indicated by the double-headed arrow. The *hatched region* around the *gal4.62* mutation was later identified as the site of mutation that gave rise to constitutive phenotype (see section 5.3.5). The total genetic length of *GAL4* locus was determined to be 0.44 cM **c** Crossing over between the two mutational sites (indicated by +) during meiosis is indicated

nique developed earlier by Thomas Hunt Morgan for determining the genetic distance between genes, which we will discuss in the next chapter.

3.1.6 Special Cases of Complementation

Intragenic complementation. We know from biochemical analysis that galactokinase is a monomeric protein, while uridyl transferase and epimerase are dimeric proteins. That is, two identical protein monomers have to associate to give a functional uridyl transferase or epimerase. In a given epimerase minus mutant haploid strain, only one type of mutant monomeric polypeptide would exist. However, in heterodiploids formed by crossing two epimerase mutant haploids bearing different alleles, the cytoplasm will have two species of mutant epimerase monomeric units unlike either of the haploids. It has been observed in certain cases that different versions of the mutant monomeric polypeptides can sometimes associate to form functional enzymes giving rise to complementation. Because of the above situation, two mutant strains otherwise belonging to the same complementation group may be classified as belonging to different complementation groups (recall the definition of complementation). This phenomenon is called "intra-genic" or "inter-allelic" complementation (Fig. 3.1.5). It should be noted that neither of the two alleles would give intragenic complementation. Only specific alleles would complement and this is known as allele-specific complementation. Intragenic complementation (note that during complementation analysis these two mutants would be assigned to different complementation group, but genetic mapping would indicate that they are alleles based on their location, see next chapter for details) between two haploid mutants, indicates direct protein–protein interaction. For example, no diploids formed between any of the 23 haploids bearing different alleles of *GAL4* showed intragenic complementation suggesting that Gal4p may function as a monomer. Alternately, since intragenic complementation is allele-specific, probably the right kinds of alleles were not represented among the 23 alleles and therefore, lack of intragenic complementation does not necessarily rule out protein–protein interaction. In fact, we will later learn that Gal4p functions as a dimmer.

Second site non-complementation. If two mutants defective in two separate loci do not complement it constitutes the phenomenon of second-site non-complementation. Consider a heterodimeric protein like tubulin, which is made up of α and β subunits coded by separate genes. A diploid strain with one defective copy of α and one defective copy of β chain would make functional tubulin and in principle, should complement. However, the concentration of the wild-type protein would be less than what is expected from a normal diploid cell. In fact, the concentration would be one-fourth the normal (Fig. 3.1.6). If the decrease in wild-type tubulin is sufficient to cause a defect, then the two unlinked mutation would not complement. If two haploid mutants with defects at different loci do not complement for a function,

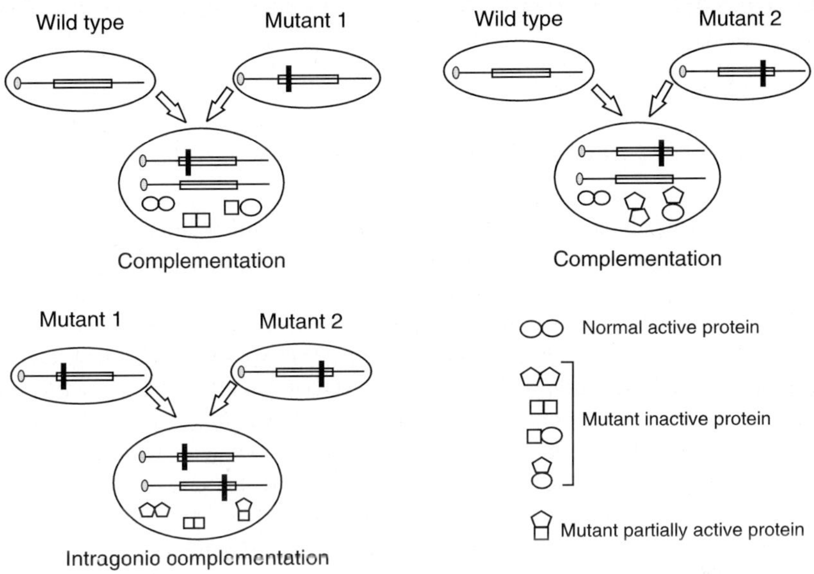

Fig. 3.1.5 Illustration of intra-genetic or intra-allelic: complementation. The *vertical bar* represents the site of mutation in two independent mutants. A diploid between mutant 1 and 2 should in principle not show complementation, but if the dimeric protein formed between the two mutant polypeptide is active, then complementation is observed. Monomeric proteins do not show intragenic complementation

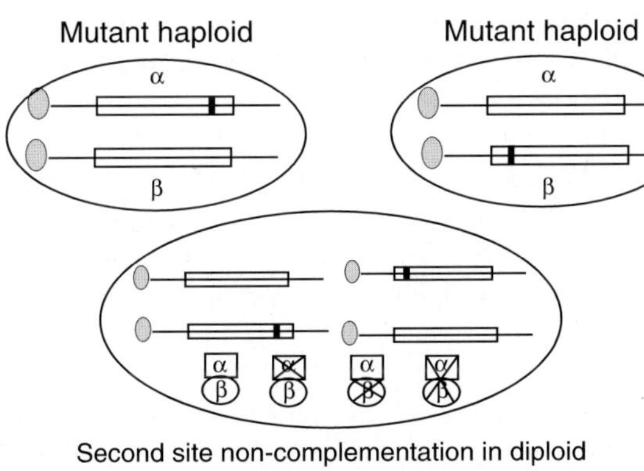

Fig. 3.1.6 Schematic illustration of second-site non-complementation. The α and β tubulin polypeptide are encoded by distinct genes present on separate chromosomes. Inactive dimers are shown by a *cross*

it indicates that the gene products interact with one another. Even this phenomenon is allele-specific.

3.1.7 *Aberrant Segregation and Recombination Model*

As previously discussed, heterodiploid formed between wild-type and a mutant defective only at one locus, yields meiotic products in a ratio of $2^+:2^-$ as determined by tetrad analysis (see segregation analysis). That the alleles segregate in $2^+:2^-$ ratio can also be inferred by random spore analysis (see Fig. 3.1.3, for details of random spore analysis). For example, if random spore analysis of *ARG4/arg 4.17* heterodiploids (1,178 asci yield 4,712 spores, see Table 3.1.3) were to be screened for arginine plus and minus phenotype, instead of $2{,}356^+:2{,}356^-$, a segregation ratio of $2{,}363^+:2{,}349^-$ (deviation from $2^+:2^-$), would have been obtained. However, in random spore analysis, this deviation from $2^+:2^-$ would be neglected due to a statistical variation (normal variation is + or $-\sqrt{n}$, where n represents the number of sample). Tetrad analysis, however, clearly indicated that this variation is not due to statistical fluctuation, but instead is due to the total of 1,178 asci 39 asci showed $3^+:1^-$ and 32 asci showed $1^-:3^+$ pattern resulting in a deviation from 2:2 pattern. Four spores of a given tetrad are sister spores, and the data of 1^+3^- or 3^+1^- segregation is interlocking and cannot be explained on the basis of statistical variation (in Fig. 3.1.2, ascus 2 shows $3^+:1^-$ segregation). In tetrad analysis of the type discussed above, a segregation ratio of 1^+3^- or 3^+1^- was consistently observed at a frequency of less than ~1.0% (Table 3.1.3). Here, one allele is unilaterally converted to the other. This phenomenon was referred to as gene conversion (Fig. 3.1.7), although one allele is getting converted to the other allele.

The departure from $2^+:2^-$ ratio discussed above is a non-reciprocal event and was initially observed in yeast and later in Neurospora. This deviation seemed to violate the law of segregation proposed by Mendel and was considered an aberration. Gene conversion was observed consistently but could not be explained based on mechanisms such as mutation. The original model that accounted for gene conversion was

Table 3.1.3 Number of asci with segregation ratios for different segregating loci

Mutant X Wt	$4^+:0^-$	$3^+:1^-$	$2^+:2^-$	$1^-:3^+$	$0^+:4^-$	Total asci
arg 4.17 X Wt	0	39	1,107	32	0	1,178
		$(117^+:39^-)$[a]	$(2{,}214^+:2{,}214^-)$	$(32^+:96^-)$		$(2{,}636^+:2{,}349^-)$
his 4 X Wt	0	143	3,546	130	0	3,819
leu 1.1 X Wt	1	40	696	24	0	760
leu 2.1 X Wt	1	19	3,676	34	0	3,729

[a]The *number in parenthesis* indicates the spores distribution with respect to the *ARG 4* locus
Wt refers to wild type

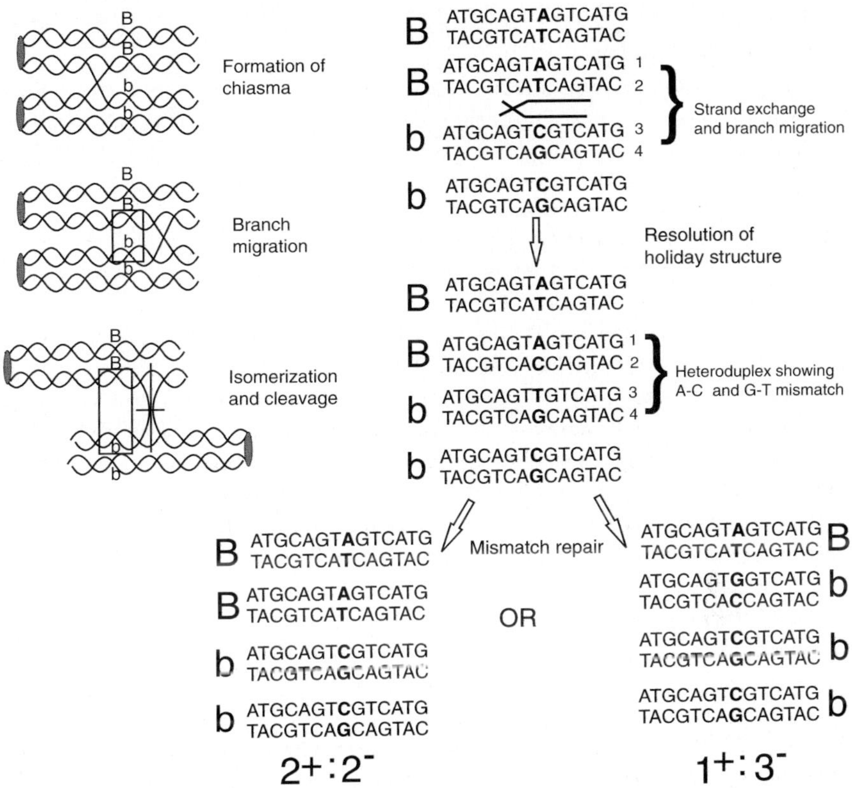

Fig. 3.1.7 Recombination model based on gene conversion event. On the *left panel*, two duplicated homologues bearing heterozygous loci represented as *B* and *b* are indicated. A nick followed by strand separation and ligation between the phosphodiester bonds of the two non-sister chromatids is indicated. The X-shaped structure migrates through the heterozygous locus leading to the formation of a heteroduplex. Following this, the structure is isomerized (as shown) and excised. This results in products having mismatched base pairs in strands 1, 2, 3, and 4, which are repaired by a mismatch repair system. In mutant allele, AT base pair is substituted with GC. The heteroduplex AC and TG can be repaired in one or the other way, resulting in gene conversion as indicated

Box 3.1.1 *Allele frequency*

Population genetics aims to study the genotypic variation existing in the population while only phenotypic variation is most easily detected. It is possible to infer the genotypic variation from the frequencies of monogenic recessive disorders. Galactosemia is a monogenic recessive disorder that occurs at a frequency of 1:40,000. This is one of the causes of juvenile cataracts, which can lead to blindness in children. The relative frequencies of different alleles of a gene in the population are called "gene frequencies", where each individual contributes two alleles. If n alleles are present for a gene in a population then $n(n+1)/2$ genotypes exist assuming random mating. Consider p as

(continued)

Box 3.1.1 (continued)

the frequency of wild-type allele and q as the frequency of mutant allele in the population. Since we are considering only two alleles $p+q=1$. Random mating without regard to genotype is mathematically equivalent to random mixing of these two alleles. Then the genotype frequencies are given by the binomial theorem

Genotype	AA	Aa	aa
Phenotype	p^2	2pq	q^2

where A and a indicate normal and wild-type alleles. q^2 (genotypic frequency) is 1/40,000, the value of q (gene frequency) would be 0.005. Since $p+q=1$, $p=1-0.005$.

From this one can calculate the carrier frequency to be approximately 0.1% (2pq). From the estimated carrier frequency it is possible to calculate the probability of a couple conceiving a baby with the disease. This is based on the principle of Hardy Weinberg equilibrium. Now that the human genome sequence is known, it is possible to directly test whether a given individual bears the defective allele, provided we know the identity of the gene responsible for the defect.

Box 3.1.2 *Recombination and complementation*

Recombination and complementation refer to the structure and function of a genetic element or a gene. In recombination, there is a physical exchange of genetic material between two chromosomes, while in complementation there is no genetic exchange. Complementation is possible because a gene elaborates a diffusible product. The following analogy brings out the distinction. Consider two copies of a 100-page book; one lacking the first ten pages and the other lacking the last ten pages. If these two books are available, one can go through the entire contents without stitching the book together. That is, the two books complement the defect. If a functional book has to emerge from these two independent defective books, then they will have to be stitched and this corresponds to recombination. Concept of complementation and recombination originated from studies carried out in fungi and drosophila. Seymour Benzer using the bacteriophage as a genetic system carried out a high-resolution recombinational analysis and coined the term "cistron", which refers to the segment of DNA that codes for a diffusible product.

Complementation analysis can be more complex than what is discussed in this section. We now know that proteins are modular and made up of independent domains that can be brought together to constitute a functional protein: for example, α complementation in β-galactosidase of *E. coli*. Here, a β-galactosidase

(continued)

Box 3.1.2 (continued)

lacking the internal codons from 21 to 41 is totally devoid of activity. This defect can be compensated in *trans* both in vitro and in vivo by a polypeptide product of first 60 codons of β-galactosidase. In addition to multidomain proteins, multifunctional proteins are commonly encountered. Complementation analysis of mutant strains bearing mutations in one gene but affecting separate functions can give rather unusual complementation patterns. Bifunctionality of proteins will be discussed with specific reference to yeast galactokinase and epimerase (see sections 6.1.2 and 8.3.3).

proposed by R. Holiday. Mismatch repair of the heteroduplex DNA formed during the recombination is a fundamental feature of this model (Fig. 3.1.7). This basic model has been refined and molecular details of this process have been well understood.

References

Bassel J, Mortimer R (1971) Genetic order of the galactose structural genes in *Saccharomyces cerevisiae*. J Bacteriol 108:179–183

Douglas HC, Hawthorne DC (1964) Enzymatic expression and genetic linkage of genes controlling galactose utilization in *Saccharomyces*. Genetics 19:837–844

Fincham JRS (1994) Genetic analysis. Blackwell Scientific Publications, Oxford

Griffiths AF, Miller JH, Suzuki DT, Lewontin RC, Gelbert WM (1996) An introduction to genetic analysis. W.H. Freeman and Company, New York

Griffiths JD, Harris LD (1988) DNA strand exchanges. CRC Crit Rev Biochem Suppl 23:43–86

Hawthorne DC (1993) *Saccharomyces* studies 1950–1960. In: Hall MN, Linder P (eds) The early days of yeast genetics. Cold Spring Laboratory Press, Woodbury, NY

Hawthorne D, Condie F (1954) The genetic control of galactose utilization in *Saccharomyces cerevisiae*. J Bacteriol 68:662–670

Matsumuto K, Adachi Y, Toh EA, Oshima Y (1980) Function of positive regulatory gene *gal4* in the synthesis of galactose pathway enzymes in *Saccharomyces cerevisiae*: evidence that the *GAL81* region codes for the part of the gal4 protein. J Bacteriol 141:508–527

Mortimer RK, Hawthorne DC (1964) Genetic mapping in yeast, Chap. 12, Meth Cell Biol 11:221–233

Holliday R (1964) A mechanism for gene conversion in fungi. Genet Res Camb 5:282–304

Holiday R (1974) Molecular aspects of genetic exchange and gene conversion. Genetics 78:273–287

Ott J (1991) Analysis of human genetic linkage. The Johns Hopkins University Press

Stahl FW (1989) Genetic recombination. Sci Am 256:91–101

Szostak J, Orr-Weaver TL, Rothstein RJ, Stahl FW (1983) The double-strand-break-repair model for recombination. Cell 33:25–35

3.2 Genetic Mapping of *GAL* Genes

3.2.1 Introduction

Genetic mapping locates the relative position of any genetic element with respect to a uniquely identifiable phenotype or a genetic marker. In genetic mapping, the frequency with which any two genes or genetic elements remain together during meiosis is measured. Even in organisms where meiosis is not a part of their life cycle, the tendency of genes to remain together is what is measured. This tendency is proportional to the genetic distance between genes or genetic elements. In fact, the technique of genetic mapping was originally intended to reflect the physical distance between genetic elements. This technique was later refined so much so that the fine structure of a gene could be determined.

The way we analyze genetic processes has dramatically changed because of our ability to sequence the whole genome. Nevertheless, the knowledge of basic concepts of genetic mapping is fundamental for understanding the genetic basis of many biological phenomena. For example, genetic mapping is the only way to identify the gene responsible for the phenotype when the biochemical basis of a disease phenotype is not understood. Yeast is an ideal organism to help understand the basic concepts of genetic mapping. Although mapping of genes is possible by both meiotic and mitotic mapping techniques, I shall focus on meiotic mapping techniques since it has been widely used to develop the genetic map of yeast. From the previous chapter, we learned that seven distinct genetic loci are required for galactose utilization. We wish to know their relative position and their distance from the respective centromere and from other known genes.

3.2.2 Tetrad Analysis

In random spore analysis, the frequency of recombination is expressed as the percentage of recombinants present in a pool of haploid spores, which is a reflection of the distance between the genetic elements under consideration. In tetrad analysis, the genetic constitution of sister spores (meiotic products of a single meiotic event) is determined to infer whether meiotic recombination has occurred in the meiosis that produced the tetrad. This technique is possible only in organisms where all the four products of a meiosis can be recovered and their phenotype detected. Tetrad analysis data is internally consistent and makes the inference more direct. We already learned from the previous chapter how segregation analysis carried out through tetrad analysis revealed the aberrant segregation which otherwise would have escaped the attention of the investigators. In addition, tetrad analysis also allows one to use centromere as a marker. That is, the position of a gene can be ascertained with respect to the centromere.

Haploid mutants bearing defective genes whose position with respect to each other is to be determined are mated to produce a heterozygous diploid. A diploid heterozygous at two loci yields only three types of tetrads. Parental ditype (PD, two spores are similar to one of the parents, the other two are similar to the second parent), non-parental ditype (NPD, two types of spores both of which are not similar to the parent) and tetratype (T, four spores will be genetically different from one another). Relative frequency of these classes of tetrads is a function of the position of the genes under consideration (see Fig. 3.2.1). An excess of PD as compared to NPD asci indicates linkage between the genes under consideration. If both genes under consideration are on different chromosomes and at least one of them is not centromere linked, or if they are widely separated on the same chromosomes then the frequency of PD: NPD:T would be 1:1:4. If both the genes are on separate chromosomes and are linked to their respective centromere, the proportion of the T asci will be reduced (Table

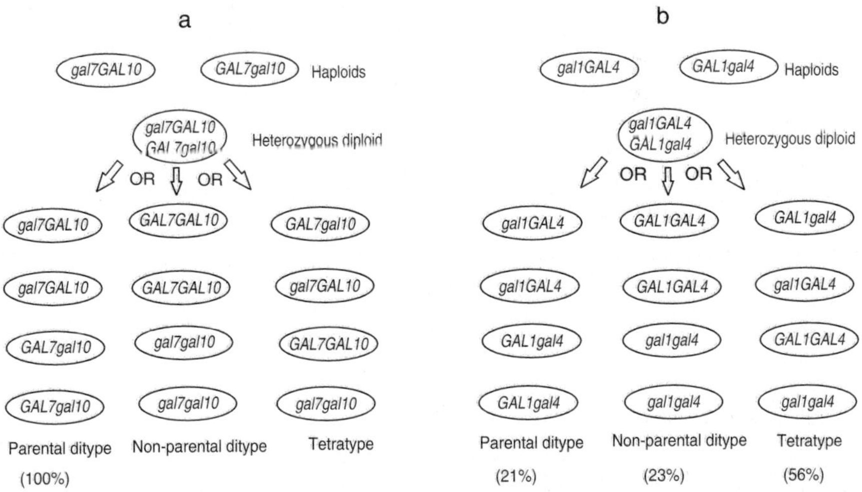

Fig. 3.2.1 Illustration of tetrad analysis. **a** Haploids *gal7* and *gal10* mutants are mated to produce the heterozygous diploid. After sporulation of the diploid, the spores are separated and their phenotype determined. The ascus is classified as either PD or NPD or T based on the genetic constitution of the spores with respect to the input haploids. In this example, 100% of asci were of PD. **b** In this example, all three types of asci were present. See Table 3.2.3 and text for details

Table 3.2.1 Distribution of class of asci obtained from double heterozygous strain

	Parental ditype	Non-parental ditype	Tetratype
Random assortment	1 (16.6%)	1 (16.6%)	4 (66.6%)
Linkage	>1	<1	4
Centromere linkage	1	1	T<4

Table 3.2.2 Distribution of tetrad classes as a function of number of exchanges

Tetrad types	Number of exchanges between linked loci							
	0	1	2	3	4	5	∞	N
PD	1	0	1/4	1/8	3/16	5/32	1/6	$1/6+1/3(-1/2)^N$
NPD	0	0	1/4	1/8	3/16	5/32	1/6	$1/6+1/3(-1/2)^N$
T	0	1	2/4	6/8	10/16	22/32	4/6	$2/3-2/3(-1/2)^N$

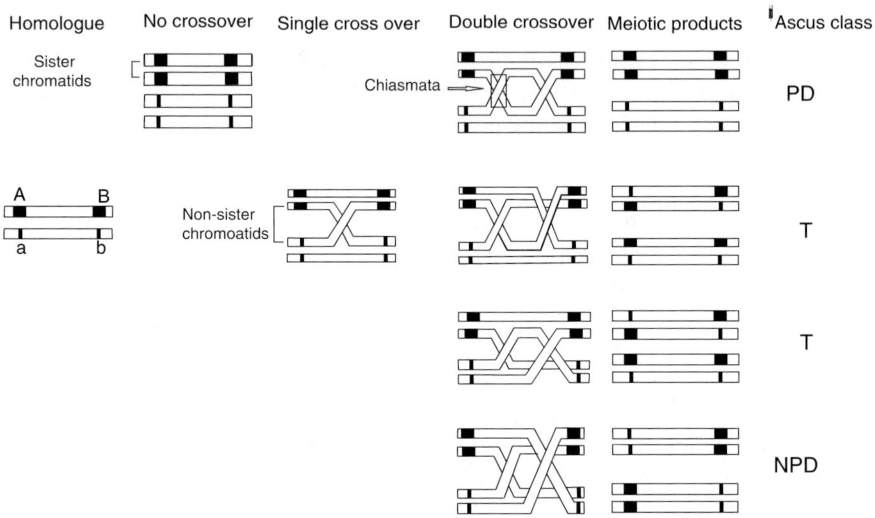

Fig. 3.2.2 Illustration of various types of tetrads with 0, 1, and 2 crossover events

3.2.1). Table 3.2.2 gives the relative mean frequencies of tetrad types as a function of the increasing number of crossing over between linked loci. When two genes are linked, the production of these three classes of tetrads when there is zero, one or two crossover events between two loci, is indicated in Fig. 3.2.2.

3.2.3 Mapping of GAL Genes by Tetrad Analysis

Only PD asci were obtained from heterodiploids formed between *gal1* and *gal7*, *gal7* and *gal10* as well as *gal10* and *gal1* (Table 3.2.3). If no crossover occurs between two loci, only PD asci are generated (Fig. 3.2.2). Note that PD asci can also be obtained by two strand even number of crossing over (Fig. 3.2.2), but then, if more than one crossing over is possible between two loci, asci belonging to NPD or T class would also be observed, which is not what is observed in the above

Table 3.2.3 Distribution of tetrad types of heterodiploids formed from haploids bearing mutation in different *GAL* genes (data obtained with permission from Douglas and Hawthorne 1964)

	Gene pair	Parental ditype	Non-parental ditype	Tetratype	Asci analyzed
1	*gal1-gal7*	313 **(100%)**	0	0	313
2	*gal1-gal10*	59	0	0	59
3	*gal7-gal10*	72	0	0	72
4	*gal1-gal2*	38 **(17%)**	39 **(18%)**	138 **(64%)**	215
5	*gal1-gal3*	27 **(41%)**	26 **(40%)**	12 **(18%)**	65
6	*gal1-gal4*	21	23	56	100
7	*gal2-gal5*	8	7	25	40
8	*gal3-gal4*	20	13	48	71
9	*gal3-trp1*	123 **(99.1%)**	0	1 **(0.8%)**	124
10	*gal2-gal3*	10	7	28 **(66%)**	45

example. Observation that only PD asci at the exclusion of the other types suggests that these three loci are tightly linked.

Heterodiploids formed between *gal1* and *gal2* showed equal PD and NPD (Table 3.2.3). When PD is equal to NPD, the two genes could be on two different chromosomes or on the same chromosome but far apart. This is equivalent to 50% recombination in random spore analysis (see below for the relationship between map distance and the RF). The tetratype asci constitute 64%, indicating that at least one of the loci is not close to its centromere. Tetratype asci gives rise to spores of which two are recombinants and two are non-recombinants (that is 50% recombination at the level of total spores) and therefore do not ascertain linkage. However, the frequency of T asci can be of value in determining the centromere linkage. If PD: NPD is 1:1, then the proportion of tetratype asci can take any value up to a maximum of 66%, depending upon the distance of the loci from the centromere. Consider the example between *gal1* and *gal3*, wherein PD equals NPD, but the tetratype is much lower than the previous example, indicating that they are closer to their respective centromere (see below). Note that all the four spores of NPD class are recombinants and this can occur without a physical exchange between chromosomes. That is, independent assortment can also give rise to NPD asci (Fig. 3.2.3). Therefore, meiotic recombination is a process that generates a meiotic product whose genetic constitution is different from that of the input haploids.

The following example illustrates how the distance of a locus from centromere determines the proportion of the tetratype asci. Recall that the centromeres of homologous chromosomes segregate with the first division of the meiosis. In the second division of meiosis the centromere splits, and the two chromatids are distributed to the sister spores. Because of this, if genes are closely linked to the centromere, no crossover occurs between the loci and centromere, resulting in low frequency of tetratype. Consider the distribution of asci with respect to *GAL3* and *GAL2* (Fig. 3.2.3). If *GAL2* were to be closely linked to the centromere, then the proportion of T asci would have been far less, which is not what is observed (Table 3.2.3). In fact, with respect to *GAL1* and *GAL3*, the tetratype is less than expected.

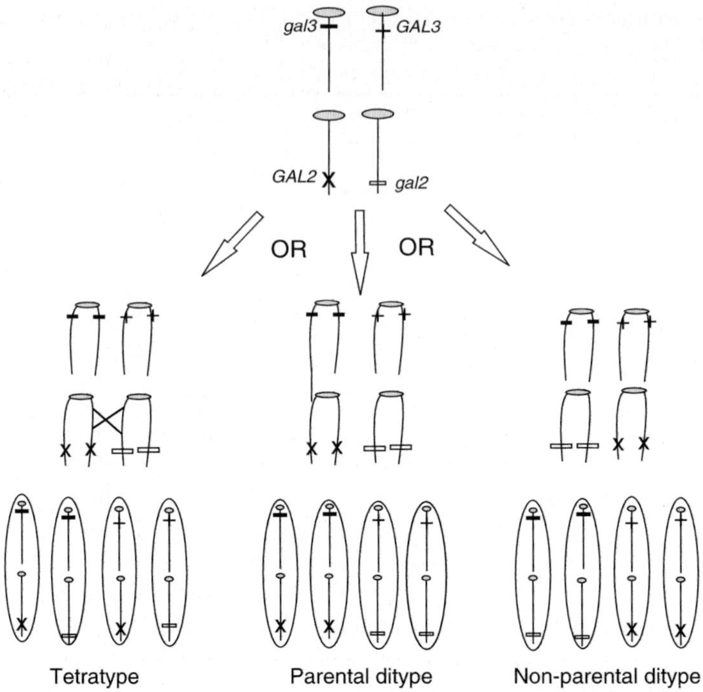

Fig. 3.2.3 Segregation pattern of *GAL2* and *GAL3* located on two different chromosomes. Tetratype asci would be generated only if a crossover occurs between the locus and the centromere. Since *GAL3* is closely linked to the centromere, no opportunity for crossing over occurs between *GAL3* and the centromere. However, crossing over can occur between *GAL2* locus and the centromere resulting in T asci. The proportion of T asci would therefore depend on the distance between the centromere and *GAL2*. If both the loci are close to centromere then the T asci will be significantly less

From these data one can infer that *GAL1* is closer to its centromere than *GAL2* is to its centromere.

The following example illustrates that genetic distances between genes involved in any biological processes can be determined provided a phenotype to score for the recombinants is available. For example, *TRP1* is one of the genes required for tryptophan biosynthesis. Haploid strain defective in *TRP1* does not grow on a medium lacking tryptophan. Heterodiploids were formed between haploid *gal3* and *trp1* mutant strains, sporulated, and the phenotypes of the spore with respect to growth on galactose and tryptophan was independently monitored to obtain the results presented in Table 3.2.3. The observation that NPD and T asci is far less as compared to PD suggest that they are linked to one anther and both of them are linked to their centromere. That certain genes are closely linked to the centromere proved helpful in cloning the centromere by using the principle of positional cloning (see section 5.5.6).

Recombination frequency, map distance (Box 3.2.1) and the distance from the centromere (if a centromere linked marker is included in the cross) can be calculated from tetrad data using the following relationship.

Box 3.2.1 *Map distance*

One genetic map unit is the distance between any two loci that yields on average 1% recombinant chromosomes or gametes. Put another way, the distance between gene pairs for which one product of meiosis out of 100 is a recombinant. Crossover frequency or recombination frequency of 0.01 or 1% is defined as one map unit or 1 cM. Since the map distance is additive, it was inferred that genes have to be located in a linear order. This conclusion was arrived at even before the physical nature of the gene was understood.

Crossing over: In the first stage of meiosis, the homologues synapse and divide lengthwise except at the centromere. During this process, crossing over can occur anywhere along the length of non-sister chromatids. This is a chance event. Chiasmata is an intermediate of the crossing-over event. Only when a crossing over occurs between the heterozygous gene loci under consideration will recombination be detected and half the meiotic products of that tetrad will be recombinants. Therefore, the chiasma frequency is twice the frequency of crossover products or recombinants. Chiasma % = 2 (% recombinants) or % recombinants = 1/2 (chiasma %).

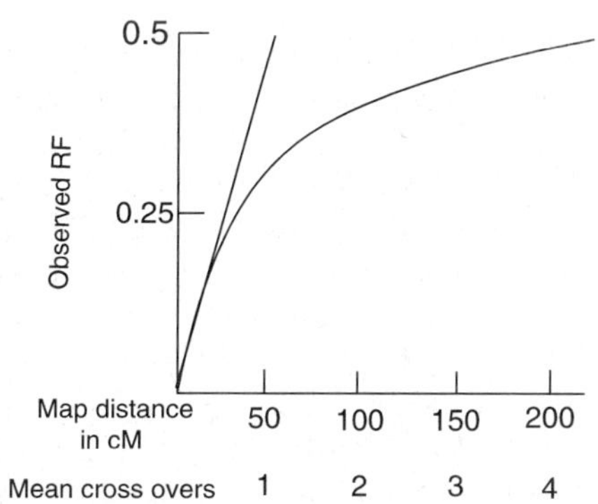

Fig. 3.2.4 Recombination frequency (RF) as a function of map distance. The *Curve* represents the relationship between the observed recombination frequency and the additive map distance. Additive map distance is arrived at by the summation of the map distances obtained from two factor crosses in which pairs of alleles span short intervals. The *straight line* shows the increase in RF as a linear function of map distance. Mean chiasma frequency can be calculated from recombination frequency by using the mapping function as discussed in the text

$$\text{Recombination frequency} = \frac{\text{NPD} + \frac{1}{2}\,\text{TT}}{\text{Total number of tetrads}}$$

Distance between two genes in cM = TT + 6NPD/2(PD + NPD + TT) × 100
Distance of a gene in cM from the centromere = TT/2(PD + NPD + TT) × 100

3.2.4 Map Distance and Recombination Frequency

No matter how far two genes are from each other, the recombination frequency never exceeds 50% (Fig. 3.2.4). That is, when two genes are far apart, 16.6% PD (this class has 0% recombinants), 16.6% NPD (100% of this class are recombinants) and

66.6% T (only 50% of this class are recombinants), would yield no more than 50% recombinants. It appears paradoxical why genes far apart on the same or on separate chromosomes can only give a recombination frequency of 50% and not more.

The recombination fraction as a function of map distance increases linearly not beyond a map distance of say 20 cM (Fig. 3.2.4). Consider linearly located loci A, B, and C with recombination frequency of 15% between A and B and 20% between B and C. Experimental determination of recombination frequency between A and C is expected to yield a frequency of 35%, which is never observed. Why is this so? Let us consider six meiosis as shown in Fig. 3.2.2. There are a total of nine exchange events between loci A and B, out of six meiosis. That is, a total of ten recombinants out of 24 meiotic products, which translates to a map distance of 41 cM, but this is an underestimation of the map distance for the following reasons: intuitively we know that the map distance is linearly related to the exchange frequency, i.e., more exchanges occurring more the distance. However, more the exchanges occurring between two loci, there is a tendency to decrease the recombination frequency. For example, two strands with a double crossover would yield no recombinants at all, although the physical event would have occurred.

Therefore, it is the frequency of the mean exchange events that is linearly related to the map distance. One can calculate the mean exchange frequency from the above data as nine events per six meiotic events, which is a mean exchange frequency of 1.5. We already know that an exchange frequency of 1 is equivalent to 50 cM, therefore the map distance is 50 times the mean exchange frequency, which is 75 cM in the above example. Therefore, the recombination frequency underestimates the distance between loci that are more than 20 cM, beyond which the linearity between RF and map distance is not observed. Imagine that in Fig. 3.2.2 the distance between loci A and B is too small and only one crossing over event has occurred in six meiosis, instead of nine. That is, two recombinants out of 24 meiotic products, which is a recombination frequency of 8%. The mean crossing over is 0.16, which translates into a map distance of 8 cM. It is clear that as the distance increases the mean exchange frequency increases proportionally, but not the recombination frequency. Therefore, that the map distances are proportional to the mean frequency of exchange rather than the probability of at least one exchange occurring. The genetic distance is directly proportional to mean number of crossing over with a proportionality constant of 50.

Unfortunately, experimentally, it is not possible to calculate the mean number of exchanges occurring between any two loci. The parameter that can experimentally be determined is the RF. Experimental approach to overcome this non-linearity between RF and map distance is to include markers close by so that the experimentally determined RF does not exceed 20%. This is not always possible. This problem is overcome through the mapping function, which states a relationship between RF and the mean crossover frequency. According to this, $RF = ½(1 - e^{-m})$ where m is mean crossing over frequency. From this relationship, m can be calculated: $m = -\ln(1-2RF)$. From tetrad data, one can calculate m without using the mapping function using the relationship $m = T + 6NPD$.

3.2.5 An Aside on Mapping Human Genes by Linkage Analysis

The relationship that the mean chiasma frequency of 1 is equivalent to 50% recombination units was used to determine the genetic length of the human genome. As early as 1955, microscopic observations revealed that an average of 52 chiasma are present on human male autosomal bivalents (tetrads). According to this, the total genetic length of human male genome was calculated to be 2,600 cM. In females, the recombination rate is about 1.5 times higher than in males and the genetic length is calculated to be 3,900 cM. Based on this, a sex-averaged autosomal map length of 3,300 cM was arrived at. While the genetic map distance is directly proportional to the physical length, the proportionality constant not only varies between males and females it can also vary between chromosomes and in different parts of the same chromosomes (Fig. 3.2.5). This is because recombination propensity depends on the nature of sequence. On average, 1 cM is equivalent to 10^7 bp in case of humans, while in yeast it is 10^6 bp.

In humans, as in yeast, linkage between genes is ascertained by calculating the recombination frequency. However, unlike in yeast, neither meiotic products nor the tetrads are available for determining the recombination frequency. This difficulty is overcome by conducting the pedigree analysis wherein the phenotype of the offspring, and parents (and the grandparents if available), is used to determine whether an offspring is a recombinant or not. Due to ethical reasons, genetic experiments cannot be conducted and therefore one depends on the available data from the population.

The classical linkage analysis can only be carried out with respect to Mendelian genes that are characterized by one-to-one correspondence between genotype at single locus and the phenotype. The success of this method depends on the availability of multigenerational families with large sib ships and polymorphic loci that are used as markers. If a disease locus is inherited in a Mendelian fashion and its biochemical basis is unknown, linkage analysis is the method of choice to locate the position of the corresponding gene. The following examples illustrate these principles.

In humans, X chromosome was first mapped using linkage analysis owing to the fact that genes present on the X chromosome can be easily identified by simple

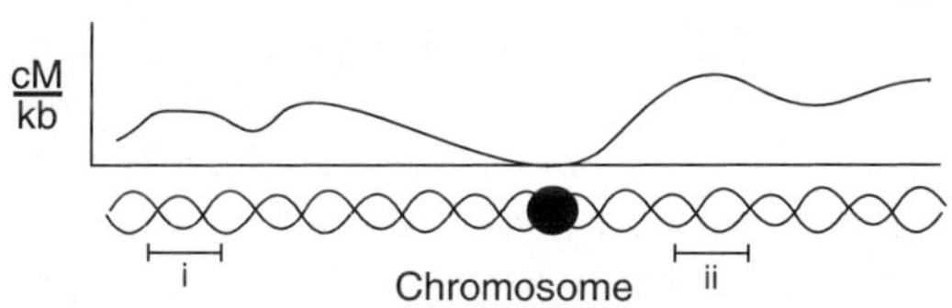

Fig. 3.2.5 Schematic relationship between genetic and physical length along the length of a chromosome. The ratio between genetic distance per unit physical distance (cm/Kb) is plotted along the length of the chromosome. The recombination frequency between any two points of equal physical distance need not be constant, as indicated between region *i* and *ii*

pedigree analysis. For example, it was earlier established that gene coding for glucose 6-phosphare dehydrogenase and the gene responsible for color blindness are present on the X chromosome, indicating that they are linked, but how far apart they are was not known. The recessive mutations in these two genes occur at a high frequency in the Black population of northern America. In one of the studies, 135 children were identified with color blindness out of a total of 3,685 screened. Based on glucose 6 phosphate dehydrogenase activity of red blood cells of these individuals, it was inferred that in ten families, the glucose 6-phosphate dehydrogenase gene was also segregating. It turned out that one out of 20 sons showed recombination, giving a value of 0.05% (Fig. 3.2.6).

Mapping of autosomal loci began in the 1950s. The detection of linkage of any two autosomal loci was more difficult due to the limitation of marker loci. Until the 1980s, polymorphic loci such as blood groups and serum enzymes were commonly used as genetic markers. For example, the gene for cystic fibrosis was shown to be closely linked to the PON locus (Box 3.2.2). Cystic fibrosis is the most common lethal genetic disorder inherited as an autosomal recessive trait and occurs in approximately one out of 2,000 live births in the Caucasian population. Increased sweat electrolyte is the diagnostic test, but the underlying biochemical basis was not understood until late 1980s. That is, the defective protein that is responsible for the

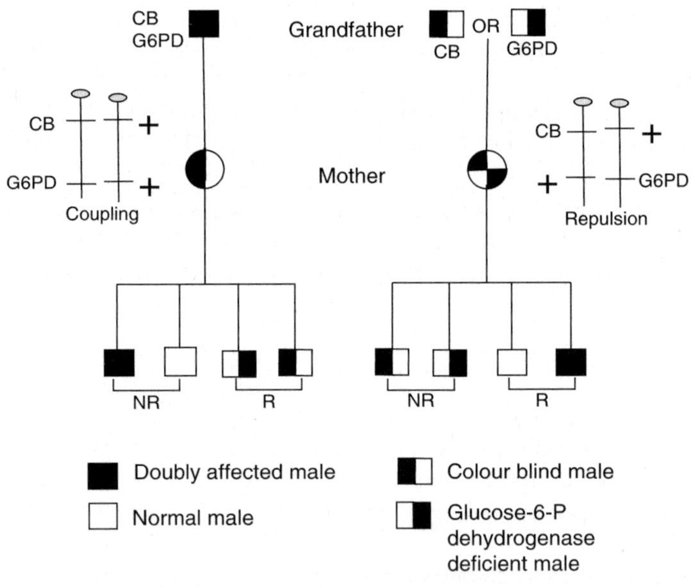

Fig. 3.2.6 Mapping of loci on X chromosome corresponding to color blindness and glucose 6-phsphapte dehydrogenase. Whether the son is a recombinant can be directly inferred by observing the phenotype of the sons of a double heterozygous mother provided the phase (coupling if the two recessive genes are on the same and repulsion if they are on different-homologous chromosome) of the linkage in the mother is known. The phase of the linkage can be ascertained based on the phenotype of the grandparents. *R* and *NR* represent recombinant and non-recombinants and + indicates wild type alleles (adapted with permission from Mckusick and Ruddle 1977)

Box 3.2.2 *Protein polymorphism*

In the 1960s, the use of serum enzyme polymorphism became a common tool to detect linkage. For example, three alleles of PON locus encoding serum arylesterase were present in the population. According to this, in any given population, six different genotypes can exist, which can be identified by subjecting the serum to native gel electrophoresis (non-denaturing, so that the enzyme activity is not destroyed) followed by detecting the presence of the enzyme by activity staining. A typical pattern indicating the six different genotypes is indicated. It should be kept in mind that the difference in electrophoretic mobility is due to a difference in the charge/mass ratio of three proteins encoded by three alleles. Genetic change need not necessarily result in alleles giving rise to proteins with different charge/mass ratio in which case the above method cannot be used to identify the alleles.

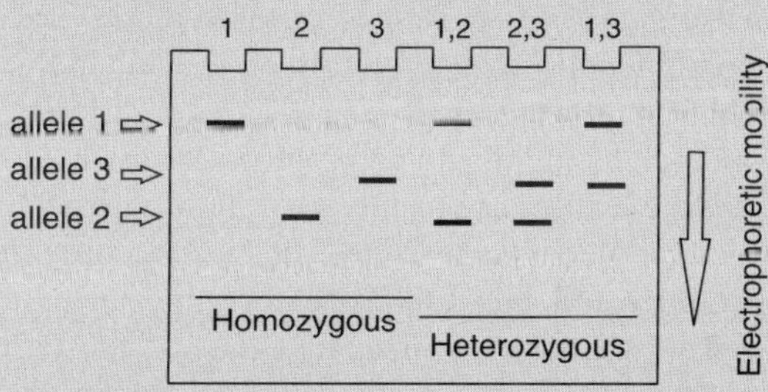

Fig. 3.2.7 Schematic illustration of protein polymorphism as detected by electrophoresis. Serum samples of individuals are loaded in separate lanes and electrophoresed under non-denaturing conditions. The position of the enzyme is detected by activity staining. Based on the pattern, it can be inferred whether the individual is homozygous or heterozygous for the locus under consideration

observed phenotypes could not be identified. The only alternative was to genetically identify the position of the defective gene with reference to any known marker. Once the position of the gene was identified, it was possible to isolate the gene using a positional cloning approach (see section 5.5.6). Here we shall focus on the genetic mapping of the cystic fibrosis gene with reference to a known marker locus.

A large number of available markers spread across the genome were tested for possible linkage with cystic fibrosis without success. Nevertheless, these data indicated that cystic fibrosis may not be linked to any one of the tested markers, thus excluding a significant portion of the genome. Eventually, PON, a locus coding for paraoxonase, a serum arylesterase (see Box 3.2.2), showed linkage to cystic fibrosis locus. Cystic fibrosis patients identified based on the clinical symptoms and the presence of positive clinical test were typed by ascertaining their genotype with respect to the PON locus.

If the PON and CF loci are truly linked with recombination fraction of θ, the likelihood of the occurrence of non-recombinant is $1-\theta$ and the likelihood of being recombinant is θ. In testing for linkage using the statistical method one considers a null hypothesis, that is, the two loci are not linked. In this case, the recombination fraction takes only one value, i.e., $\theta=0.5$. For an alternate hypothesis, that is where they are linked, the value of recombination fraction θ can vary from 0 to 0.5. The likelihood ratio, the so-called Lod score, (logarithm of odd) indicates that for a given value of $\theta<1/2$ how much higher the likelihood of the data in the absence of linkage: $Log_{10}=L_\theta/L_{1/2}$. It is the ratio of support for linkage at various values of θ divided by no linkage at $\theta=1/2$.

In the pedigree shown in Fig. 3.2.8, the combination of allele in the father is not informative, because even if a recombination occurs between the cystic fibrosis and the PON locus, it cannot be detected. That is, the father is not informative as he is homozygous with respect to PON, the marker locus. As the mother is heterozygous with respect to CF and the PON, her allelic combination, that is the phase, between CF and the PON locus can be either of two possibilities. Considering possibility I, children 1 and 3 are recombinants with a likelihood of θ and individual 2 is a non-recombinant with a likelihood of 1−θ. According to the second possibility, individuals 1 and 3 are non-recombinants with a possibility of 1−θ and individual 2 is a recombinant with the probability of θ. If the two genes are on different homologues then the probability of an individual being a recombinant or not is ½.

The probability of the occurrence of the above family structure is calculated by taking the ratio of alternate hypothesis by substituting different values of θ in $½[\theta^2(1-\theta)]+1/2[\theta(1-\theta)^2]$ and the null hypothesis by substituting the value of $\theta=0.5$ in $½[\theta^2(1-\theta)]+1/2[\theta(1-\theta)^2]$. The first part of the term in the alternate hypothesis considers that the two loci are in coupling and the second part considers that they are in repulsion configuration, both of which are equally likely and average value is taken. The logarithm of the ratio of the above probabilities gives the Lod score (Z). The use of logarithms allows data collected from different families to be combined by addition. A positive value of Z suggests that the loci under consideration are linked, whereas negative values suggest that linkage is less likely than the possibility that the two loci are unlinked.

In the second pedigree, the phase of the linkage is known because of the availability of the grandparental information and the offspring can be identified as either a recombinant or not. Because of this, the second term of the alternate hypothesis in the above equation is no longer required. Table 3.2.4 gives the Lod score for different values of θ for the above two pedigrees. The most likely recombination fraction is the value of θ, which gives the highest positive Lod score.

According to the above data, the first family does not ascertain linkage between PON and cystic fibrosis. In the second family, the probability of these two loci being linked at $\theta=0.4$ is twice that compared with no linkage. A maximum Lod score of 3 (or an odds ratio of 1,000 to 1) is required to assert linkage. This high stringency of an odds ratio of 1,000:1 is called for due to the low prior probability that any two randomly selected loci are linked. That is, it is inherently unlikely that any two loci chosen randomly would be present on the same chromosome out of a possible distribution into 22 autosomes. Stronger evidence is required if something is inherently improbable. The odds ratio of 1,000:1 ensures that the probability of

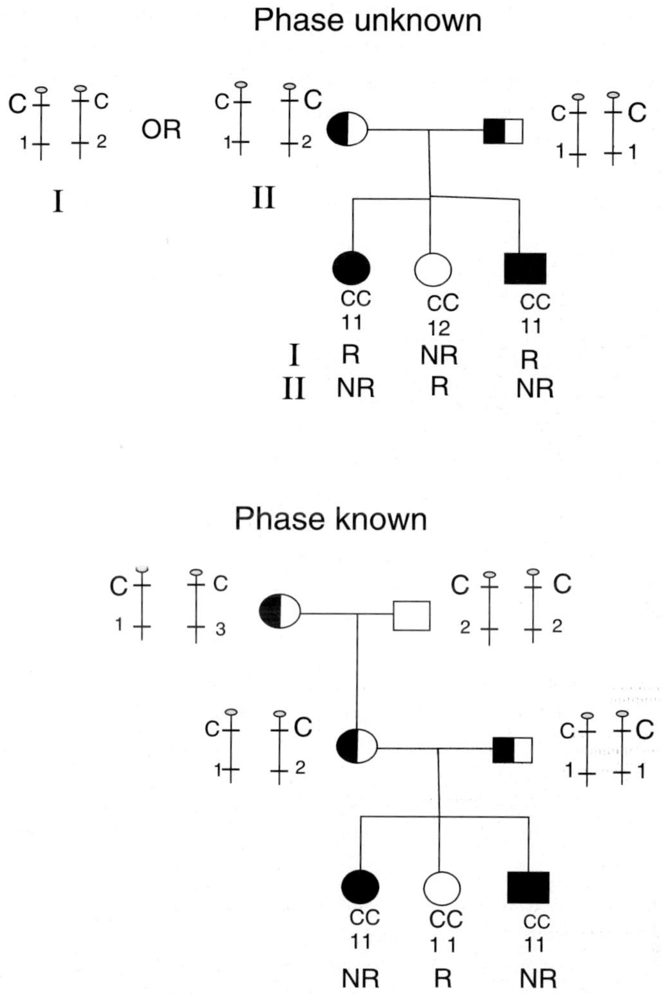

Fig. 3.2.8 Human linkage analysis. A typical pedigree of a two- and three-generation family with one normal and two children affected with cystic fibrosis is shown. The relevant genetic constitution with respect to cystic fibrosis and PON locus is represented. *1* and *2* represent different alleles of the PON locus while c and C represent mutant and wild type alleles of cystic fibrosis gene respectively. The parents are normal but are carriers of the recessive disorder. If the two loci are un-linked they will be present on two separate homologues

Table 3.2.4 Lod scores for the two pedigrees shown in Fig. 3.2.8

θ	0.05	0.1	0.2	0.3	0.4
Z (Phase unknown)	−1.0	−0.443	−0.193	−.077	−0.017
Z (Phase known)	−0.442	−0.188	0.010	0.070	0.311

a false-positive is less than 5%, which is equivalent to a p-value of 0.05. In general, families are analyzed until a positive Lod score of at least 3 is obtained.

A total of 41 two-generational families with at least two affected children were considered. Serum samples were analyzed from affected as well as parents. The

Table 3.2.5 Lod scores (Z) between PON and cystic fibrosis at various recombination fractions (data obtained with permission from Schmiegelow et al. 1986)

Recombination fraction	0.05	0.10	0.15	0.20	0.30	0.40
Lod score	2.93	3.38	3.12	2.61	1.40	0.40

Box 3.2.3 *Genomic imprinting and human diseases*

In addition to the law of segregation and independent assortment, Mendel discovered the principle of equivalence in reciprocal crosses. That is, regardless of which parent contributes a gene to their offspring, the gene will behave in the same way. It is now known that this principle is not always true; that is, it does matter from which parent the gene is inherited. Non-equivalence due to reciprocal crosses in itself is not something new. For example, the inheritance of X-linked recessive traits is non-equivalent in reciprocal crosses. The second example of non-equivalence involves traits controlled by non-nuclear genes. For example, mitochondrial encephalomyopathy is one example. The above examples are due to unequal genetic contributions by each parent and therefore cannot be considered as reciprocal crosses in a strict sense. In genomic imprinting, parents make exactly identical genetic contributions, and yet defy the principle of equivalence. In general, we assume that both the paternal and maternal alleles behave the same way unless they are mutated. It was initially believed that this phenomenon is rare. Monoallelic expression of biallelic genes is now known to occur in many cases. In general, monoallelic expression of biallelic genes is independent of parental origin as in inactivation of X chromosome or allelic exclusion in immunoglobulin gene. This occurs without regard to the parental origin. If monoallelic expression is dependent on parental origin then it is referred to as genomic imprinting.

Prader-Willi and Angelman syndromes are due to problems with differentially imprinted genes at 15q11-q13. The former is characterized by mental retardation, hypotonia, gross obesity male hypogenitalism. While the latter is characterized by mental retardation, inappropriate laughter, and hyperactivity. The individual will suffer from PWS if he does not receive at least a copy of gene present in 15q11-13 from the father no matter if the individual receives two normal copies of this region from the mother. In contrast, an individual will suffer from AS if he does not receive at least one copy of this region from his mother no matter if he receives two normal copies from the father. One or more genes present in this region are erased of their previous parental history and imprinted during gametogenesis. Paternal and maternal deletions in 15q11-q13 region are responsible 70% of cases of PWS and AS, respectively.

Box 3.2.4 *Polygenic traits*

Most genetic disorders that have been well understood are inherited in a Mendelian fashion. In reality, a large number of disorders are due to multiple genes; such types of disorders are called "multifactorial" or "polygenic". Identifying the contribution of each gene for such traits is difficult. In addition, environmental factors also contribute, making the analysis that much more complicated. Attempts are underway to analyze the genetic basis of such disorders.

Box 3.2.5 *Other uses of tetrad analysis*

Often, in genetic analysis, one requires isogenic strains with different genetic backgrounds. Using tetrad analysis it is possible to isolate haploids of a specific genotype starting from a heterozygous diploid. It is also possible to identify whether a combination of any two genetic defects lead to lethal phenotype. For example, consider two individual haploids bearing single gene defects which by themselves are not lethal. If the two defective genes in combination lead to lethality (synthetic lethality), then spore that receives these genes will not be viable, which can be easily discerned.

Box 3.2.6 *Chi-square test*

In the example given in Table 3.2.3, it is easy to conclude that *GAL1* and *GAL7* are linked. On the other hand, can we conclude that *GAL1* and *GAL3* (see Table 3.2.3) are centromere-linked because the proportion of tetratype asci is less than 66%? Is it by chance that the tetratype asci is less than the expected value of 66% and that *GAL1* and *GAL3* are not really centromere-linked? Probably, if the experiment were to be repeated, one would get a ratio of 12:12:48 (1:1:4) instead of 27:26:12 (4:4:2). In other words, how do we quantify the magnitude of variation from the expected value to arrive at a conclusion? One way to establish whether the observed result is a variation is to increase the number of tetrads for analysis, which tends to decrease the variation (if any) or repeat the experiment many times. If the observed variation is by chance then such variation will not occur very often, but many times it may not be feasible to satisfy these conditions. For example, in human genetics, we are restricted with the available data points. Under such circumstances one needs to evaluate the probability of an event occurring. That is, is what we observed due to a chance event or is it real? Such decisions are important to formulate new sets of experimental strategy. Let us hypothesize that in the above example, *GAL1* and *GAL3* are centromere-linked and that

(continued)

Box 3.2.6 (continued)

is why the tetratype asci is 18% and it is not by chance that the result has turned out the way it is. A skeptic would say that it is not true and that it is by chance that it has occurred, and *GAL1* and *GAL3* are not centromere-linked and the actual number should have been 12:12:48.

To evaluate which of the hypothesis is correct, the value of χ^2 is calculated by summing the squares of the deviations from expectations divided in each case by the expected number. The value of χ^2 has to be judged from the degree of freedom (df) which is N-1 where N is equal to the number of classes under consideration. In the above example, the df is 2.

The value of χ^2 is 63.34, with a df of 2, this corresponds to a p-value of less than 0.001. That is, the probability of the ratios obtained as they are, because of chance alone is extremely low. That is, if the experiment is repeated 10,000 times, only once a deviation as large as what is observed would occur because of chance alone. Therefore, it is unlikely that the data obtained is by chance and so it has to be real. By convention it has been agreed upon that an event occurring by chance is unlikely to occur more often than 5%. Based on this, we accept the hypothesis and agree with the optimist that *GAL1* and *GAL3* are in fact centromere-linked.

The χ^2 test is applicable only to whole number samples and not to measurements. This test depends on the assumption that the number in different classes vary symmetrically about their mean values. Therefore, as a general rule, the method is used when the sample number is above a certain value. The absolute numbers are taken for this analysis and not percentages.

Lod score was calculated as before. This analysis indicated that the PON and the locus responsible for cystic fibrosis phenotype are linked by a distance of 10 cM since the Lod score at this value of θ is maximum (Table 3.2.5), but this does not tell us in which autosome these two loci are present. We shall discuss this in section 5.5.6. Lod score analysis (except in simple cases) is entirely dependent on computer programs that consider other parameters such as allelic frequencies of the marker and the branching pattern in the pedigrees.

References

Bassel J, Mortimer RK (1971) Genetic order of the galactose structural genes in *Saccharomyces cerevisiae*. J Bacteriol 108:179–181

Chandra HS, Nanjundiah V (1990) The evolution of genomic imprinting. Dev Suppl, pp 47–53

Douglas HC, Hawthorne DC (1964) Enzymatic expression and genetic linkage of genes controlling galactose utilization in *Saccharomyces*. Genetics 49:837–844

Fincham JRS, Day PR (1971) Fungal genetics, vol. 4. In: Burnett JH (ed) Botanical monographs, vol. 4. Blackwell Science Publications, Oxford
Fincham JRS (1994) Genetic analysis. Blackwell Science Publications, Oxford
Hawthorne DC, Mortimer RK (1960) Chromosome mapping in *Saccharomyces*. Centromere-linked genes. Genetics 45:1085–1110
Holliday R (1989) A different kind of inheritance. Scientific American, pp 40–48
McKusick VA, Ruddle F (1977) The status of the gene map of the human chromosome. Science 196:390–405
Mortimer RK, Hawthorne DC (1966) Genetic mapping in *Saccharomyces*. Genetics 53:165–173
Mortimer RK, Hawthorne DC (1975) Genetic mapping in yeast. Meth Cell Biol 11:221–232
Mortimer RK, Schild D (1982) Genetic mapping in *Saccharomyces cerevisiae*. Molecular biology of *Saccharomyces*, Cold Spring Harbor Monogram, vol. 1, 11–26
Risch N (1992) Interpreting Lod scores. Science 255:803–804
Risch N (2000) Searching for genetic determinants in the new millennium. Nature 405:847–855
Sapienza C (1990) Parental imprinting of genes. Scientific American, October, pp 26–33
Schmiegelow K et al (1986) Linkage between the loci for cystic fibrosis and paraoxonase. Clin Genet 29:374–377
Sherman F (2002) Getting started with yeast. In: Guthrie C, Fink GR (eds) Methods in enzymology, vol. 350. Guide to yeast genetics and molecular and cell biology. Academic Press, New York, pp 3–41
Strachan T, Read AP (1999) Human molecular genetics. Bios Scientific Publ Ltd
White R, Lalouel J (1988) Chromosome mapping with DNA markers. Scientific American 258(2):20–28

Chapter 4
Genetic Analysis *GAL* Genetic Switch

4.1 Negative Control by the Repressor

4.1.1 Introduction

We learned from the previous chapters that in strains bearing recessive mutations in *GAL4*, Leloir enzymes are not induced in response to galactose. On the other hand, recessive mutation in *GAL3* leads to a delayed kinetics of induction of the Leloir enzyme in response to galactose. In a broad sense, *GAL4* and *GAL3* are regulators of the expression of a family of genes required for galactose utilization and such genes are called regulatory genes. The fundamental question that needs to be understood is how do Gal3p and Gal4p interact to perceive the presence of galactose in the medium? What prevents Gal4p from activating the *GAL* genes in the absence of galactose? Is there another protein that represses Gal4p from activating transcription in the absence of galactose? If so, can we identify such a gene and study how galactose inactivates this gene function? In essence, we wish to understand the strategy employed by yeast to translate the three-dimensional information of galactose into a specific biological signal.

If a negative regulator prevents *GAL4* function in the absence of galactose, then it is possible to isolate a mutant strain defective in such a repressor. Such a mutant strain would express *GAL* genes constitutively. In this case, one is looking for a strain that expresses Leloir enzymes even if galactose is not present. How do we detect the presence of such rare variants in the population? For example, if the frequency of occurrence of such mutants in a mutagenized population of cells is 10^{-3}, then, one has to screen at least 1,000 independent mutant strains for enzyme expression with the hope of finding one such mutant, which is obviously a difficult task. Developing a selection or screening strategies to identify mutant strains with novel phenotypes is at the heart of genetic analysis. These rare variants provide initial clues that help us reveal mechanisms of regulation which are otherwise difficult to discover. The following example illustrates this point.

P.J. Bhat, *Galactose Regulon of Yeast.*

4.1.2 Discovery of a Repressor of GAL Regulon

Douglas and Pelroy, by chance, observed that *gal3* strain lacking mitochondrial function (yeast strains lacking mitochondrial function are represented as δ^-; such strains can be obtained by treating cells with ethidium bromide and they cannot grow on a non-fermentable carbon source such as ethanol) does not grow on galactose as the sole carbon source (recall that a *gal3* strain bearing functional mitochondria shows LTA). This observation led them to isolate variants of *gal3* δ^- that grow on galactose. When 10^5 cells of *gal3* δ^- genotype were plated on medium containing galactose as the sole carbon source, colonies at a frequency of 2×10^{-5} appeared (Fig. 4.1.1). Further studies indicated that the ability of these mutant strains to grow on galactose is faithfully transmitted to its descendents. They concluded that in these mutant strains, either *gal3* has reverted back or a mutation at a locus other than *GAL3* overcomes the original defect of inability to grow on galactose (Fig. 4.1.1).

This is the right juncture to digress from the main line of thought to discuss the concept of reversion and suppression. In the above example, *gal3* δ^- strain recovers from the galactose growth defect due to a genetic change. If the recovery is due to a reversion of the defective *gal3* locus, the phenomenon is referred to as reversion. Reversion could be due to a back mutation giving rise to the wild-type *GAL3* allele or a mutation elsewhere in *GAL3* that compensates the defect of the original mutation. The former phenomenon is referred to as "true reversion" while the latter is called "intragenic reversion" or the "second-site reversion" (sometimes referred to pseudo- reversion). Alternatively, if the phenotype is recovered because of a mutation in gene other than *GAL3*, it is referred to as "suppression" or specifically "extragenic suppression" to distinguish from intragenic suppression. A true revertant is of no use in further genetic analysis. Both intragenic and extragenic suppressor mutations are of immense value. Of course a geneticist would love to obtain an extragenic suppressor while a biophysicist would prefer the latter. How do we distinguish between true reversion, intragenic reversion and extragenic suppression? Mutation leading to true reversion can be distinguished form intragenic suppression by carrying out fine structure analysis as discussed previously (section 3.1.5). Extragenic suppression can be distinguished from intragenic by segregation analysis as discussed below.

An unexpected feature of the mutant strain is that it constitutively expresses (expression of enzymes in the absence of galactose) all Leloir enzymes. A diploid obtained by crossing the above mutant strain to a wild-type strain did not show constitutivity but exhibited normal induction, indicating that the suppressor mutation is recessive with respect to the constitutive phenotype. Douglas and Pelroy analyzed the phenotype of haploid spores from a large number of asci obtained from the above diploid (Fig. 4.1.1). Results of the phenotypes of four spores obtained from one ascus is given in Table 4.1.1. This result suggests that most probably the phenotypic reversion is due to an extragenic suppression. If it were not so, all the four spores should have shown rapid induction phenotype, which is not what is observed. They inferred that the constitutivity is due to a mutation at a locus different from

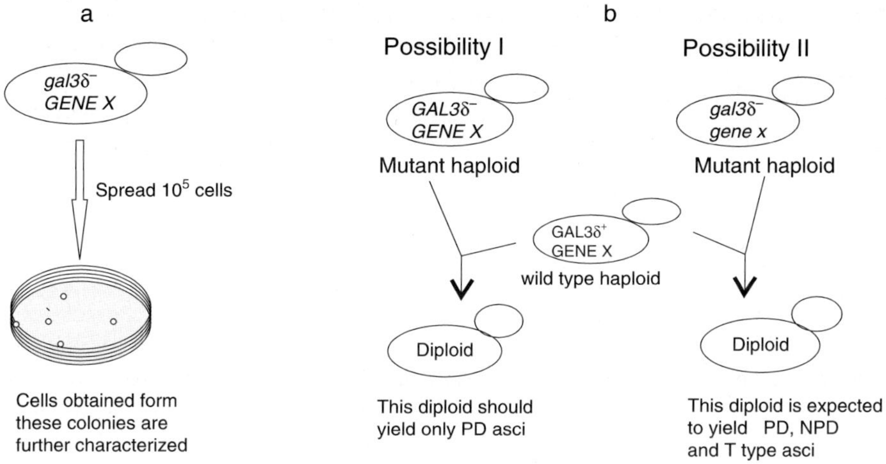

Fig. 4.1.1 Isolation of suppressors of non-inducible phenotype of *gal3 δ⁻* strain. **a** Illustrates the strategy of isolating suppressors of *gal3δ⁻* strain. **b** Possibility I considers that the reversion of mutant *gal3* to wild-type *GAL3* while possibility II considers a mutation in a gene other than *gal3*, indicated by *X*. These two possibilities have different expectations upon diploid formation with the wild-type, followed by tetrad analysis. Possibility I is expected if true reversion has occurred or extragenic suppression due to a mutation in a gene closely linked to *GAL3*. Possibility II is experted if a mutation has occured in a gene not tightly linked to *GAL3*

Table 4.1.1 Properties of segregants obtained from the diploid formed between the wild-type and a haploid mutant isolated from *gal3 δ⁻* strain (data obtained with permission from Douglas and Pelroy 1963)

Genotype			Constitutive enzyme synthesis		
GAL3 locus	*GAL80* locus	Fermentation	Kinase	Transferase	Epimerase
gal3 δ⁻[a]	*GAL80* (parent haploid strain)	–[b]	0[c]	0	0
gal3 δ⁻	*gal80* (mutant haploid)	+[b]	3.0	9.2	7.4
GAL3/gal3	*GAL80 /gal80* (diploid)	+	0	0	0
GAL3	*gal80* (spore)	+	17.0	7.8	87.0
gal3	*gal80* (spore)	+	5.2	3.7	25.2
gal3	*GAL80* (spore)	Delayed	0	0	0
GAL3	*GAL80* (spore)	+	0	0	0

[a] Indicates mitochondrial deficiency
[b] Indicates presence (+) or absence (–) of fermentation
[c] Indicates absence of enzyme activity

GAL3 and was called *GAL80*. Detailed genetic mapping indicated that *GAL80* is not linked to any of the known *GAL* genes.

Let us recapitulate the phenotypes of recessive mutations in *GAL3, GAL4* and *GAL80*. Recessive mutation in *GAL3* confers long-term adaptation phenotype, *GAL4* confers no induction even in presence of galactose whereas *GAL80* shows constitutive phenotype. On further analysis, it was observed that a recessive mutation in *GAL80* does not alleviate the need for a functional *GAL4* but alleviates the need for *GAL3* or galactose. That is, a haploid strain of the constitution *gal-80gal4* (such strains can be constructed using tetrad analysis, see previous chapter) is neither constitutive nor inducible. On the other hand, a haploid *gal3gal80* is constitutive. That is, *GAL4* is absolutely required to turn on the genes, no matter what function *GAL3, GAL80* and galactose perform. How do *GAL3* and *GAL80* in conjunction with galactose perform their roles? Do they interact with each other or with other proteins yet to be identified?

Reference

Douglas HC, Pelroy G (1963) A gene controlling inducibility of the galactose pathway enzymes in *Saccharomyces*. Biochim Biophys Acta 68:155–156

4.2 Operator Repressor Model of *GAL* Regulon

4.2.1 Introduction

Models are abstract contraptions that adequately explain observed phenomena by considering a few among many possibilities. A model need not necessarily reflect a reality but certainly makes us come closer to the reality. A model is built to test the predictions of the hypothesis and the behavior of the system under varying conditions. If the prediction turns out to be right, the model is tentatively correct, if not, it is modified to include the new insights. By now, it would have become clear, that the purpose of studying the behavior of genetic variants is to build a model that explains how a wild-type strain responds to galactose.

Based on the genetic data discussed thus far, Douglas and Hawthorne proposed that in a wild-type cell, Gal80p, the repressor, inhibits the expression of Gal4p by interacting with the operator locus of *GAL4*. Through an unknown mechanism, galactose in conjunction with Gal3p, inactivates Gal80p, the repressor. Inactivation of Gal80p results in the expression of Gal4p, which activates the transcription of Leloir genes probably by binding to the promoters (Fig. 4.2.1). Recall that a *gal3* strain takes unusually long time to induce the enzyme synthesis in response to galactose suggesting a partial defect in the inactivation of Gal80p.

4.2.2 Testing the Predictions of the Model

If the above model is correct, it is possible to isolate strains bearing mutation in the operator locus defective in interacting with Gal80p. Such a mutant strain would express Gal4p even in the absence of galactose and Leloir genes would be expressed constitutively. As predicted, constitutive mutations were isolated using the selection

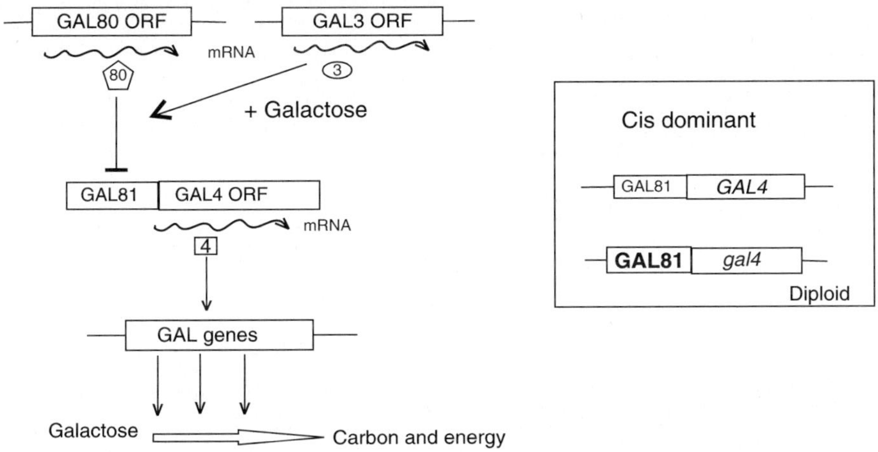

Fig. 4.2.1 Douglas-Hawthorne model of *GAL* gene activation. In a wild-type strain, in the absence of galactose, Gal80p interacts with GAL81, the operator locus of GAL4, to inhibit the expression of Gal4p. In the presence of galactose, Gal80p is inactivated, resulting in the expression of Gal4p, which in turn activates the *GAL* gene expression. *Arrow* indicates activation and a *line with a dash* indicates inhibition. A diploid with mutant *GAL81* locus (indicated in bold) is contiguous with the recessive *gal4* allele and does not show dominant constitutive phenotype. That is, the mutant *GAL81* allele can only affect the gene that is physically contiguous and does not act in *trans*. Therefore, the diploid of the constitution shown is not constitutive but is inducible

strategy described in Fig. 4.2.2. Moreover, as expected, the mutants turned out to be dominant. Preliminary genetic mapping studies indicated that the dominant mutation is tightly linked to the *GAL4* locus. Further genetic analysis indicated that the mutation is also *cis*-dominant (see Fig. 4.2.1), that is, the mutant locus does not code for any diffusible product. The wild-type operator locus was referred to as *GAL81* (Douglas and Hawthorne originally designated the wild-type locus by the letter c and the dominant allele as C. *GAL81* nomenclature is in conformity with the three-letter code).

Douglas and Hawthorne provided another piece of genetic evidence in support of the hypothesis. They argued that a *GAL80* allele insensitive to the inducer should be dominant and not turn on *GAL* genes even if galactose is present. Consistent with this, they isolated a dominant *GAL80* allele, designated as $GAL80^{s\text{-}0}$, a super-repressor, which did not respond to galactose (Fig. 4.2.2). What is the phenotype of a haploid strain bearing both $GAL80^{s\text{-}0}$ and ***GAL81***-*GAL4* allele (bold indicates dominant constitutive allele isolated by Douglas)? It turned out to be constitutive.

In genetic exploration of biology, one tries to obtain mutants that disturb the system behavior. In fact, it is the mutant phenotype that provides the clues for understanding the inner working of a cell. In modern parlance, this is called the system biology approach. The proposition of Douglas and Hawthorne on the nature of interaction between Gal80p and Gal4p was profoundly influenced by what was then known in the *Lac* system of *E. coli*. The main difference between these two was that in *E. coli* only

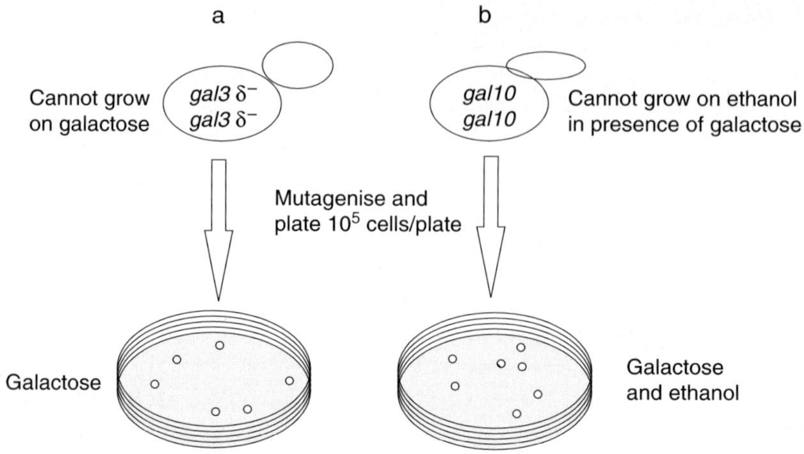

Fig. 4.2.2 Strategy for isolating dominant constitutive and non-inducible mutations in *GAL4* and *GAL80* locus, respectively. **a** A *gal3δ⁻* homodiploid, cannot grow on galactose. Mutants of this strain that can grow on galactose as the sole carbon source were sought by plating the mutagenized cells on galactose medium. The mutants thus obtained expressed *GAL* gene constitutively. The mutant diploid was sporulated and the haploid-expressing *GAL* gene constitutively was found to have a mutation linked to *GAL4* locus by mapping. **b** To isolate non-inducible dominant mutations at the *GAL80* locus, *gal10* homodiploid was used as the starting strain. This cannot grow on ethanol in the presence of galactose due to the accumulation of galactose 1-phosphate. Mutants that can grow on medium containing galactose plus ethanol were isolated. These mutants were characterized further to identify the dominant constitutive mutants and the recessive *gal10* locus was substituted with wild-type *GAL10* by genetic crosses. Diploid strain were used for isolating dominant mutations to reduce the probability of obtaining recessive constitutive mutations

negative regulation was known, while in yeast, both negative and positive regulation was noticed. That is, it is not sufficient to inactivate the *GAL80* alone, but *GAL4* has to be present if induction has to occur. Detailed genetic analysis conducted by other researchers invoked a direct protein–protein interaction between Gal4p and Gal80p (see section 4.5). Gal4p and Gal80p could not be purified as no in vitro assayable biochemical activity was assigned. These proteins could not be purified using specific tags as is routinely done at present, since the corresponding genes were not isolated. By now you would have realized the difficulty in studying regulatory proteins for which assigning a biochemical activity is not that straightforward.

References

Douglas HC, Hawthorne DC (1966) Regulation of genes controlling synthesis of the galactose pathway enzymes in yeast. Genetics 54:911–916

Douglas HC, Hawthorne DC (1972) Uninducible mutants in the *GAL* i locus of *Saccharomyces cerevisiae*. J Bacteriol 109:1139–1143

4.3 Genetic Interactions

4.3.1 Introduction

Biological manifestations are due to a network of molecular interactions governed mainly by proteins encoded by single copy of the gene as in haploids or by two copies as in diploids. Mendel invoked the concept of recessiveness and dominance to explain the consequence of loss of one or both the copies of the gene in a diploid organism. The concept of epistasis was introduced by R. Fisher in his classic paper to reconcile the opposing views of inheritance proposed by Mendel and Francis Galton. Geneticists have used these concepts extensively to unearth the underlying mechanistic and quantitative basis of molecular interactions responsible for a given phenotype. Although we say that an allele is dominant or recessive, it is the function encoded by an allele and not the allele by itself is dominant or recessive. Similarly, a function coded by a gene is epistatic over the function coded by some other gene. Often, it is the phenotypic consequence due to recessiveness, dominance, or epistasis that gives the first clue to the network of gene interactions. It is important to realize that haploid strains do exhibit recessive or dominant phenotypes, but whether an allele is dominant or recessive can only be inferred from a diploid status. However, an epistatic relationship can be inferred from haploid or diploid status.

4.3.2 Recessivity and Dominance

A diploid strain of genetic constitution *GAL1/gal1* is indistinguishable from a wild-type strain, in terms of growth on galactose. This is because the concentration of galactokinase expressed from one wild-type allele is sufficient to provide the required flux through galactose metabolic pathway. It is the flux of galactose that eventually connects activity of the enzyme to the phenotype. That is, in a heterozygote, the flux is insensitive to a decrease in galactokinase by 50% (Fig. 4.3.1). This is because the amount of enzyme present in a wild-type diploid is excess than what is required to provide the necessary flux. In other words, dominance is a mechanism that resists a decrease in the flux due to a decrease in the concentration of galactokinase. In other words, the metabolic system is robust (see section 8.5.2) to a decrease in the concentration of the enzyme.

According to the flux-versus-enzyme-activity relationship, the concentration of the enzyme in the diploid determines the extent to which the flux gets reduced as a function of the enzyme concentration. If a wild-type diploid has a limited amount of enzyme, then a decrease in enzyme levels by half due to one defective copy of the allele, would reduce the flux to give rise to a phenotype (Fig. 4.3.1). In this case, one defective allele results in a dominant phenotype. Although it is rare that a defect in one copy of the gene encoding an enzyme gives rise to a dominant phenotype, maturity onset of diabetes of the young (MODY 2) is characterized by autosomal dominant inheritance due to defective glucokinase. That is, glucokinase produced from one normal allele is not sufficient to perform its function. These individuals suffer from a variety of defects such as defects in glucose-induced insulin secretion and impaired hepatic glycogen synthesis.

Haploinsufficiency, that is the lack of a functional allele, is a common cause for a majority of human genetic disorders. Most often haploinsufficiency of regulatory proteins which function in a stoichiometric fashion give rise to dominant phenotypes even when the concentration is reduced by 50% of the normal levels.

We learned that a strain of the genotype *GAL81GAL4* /***GAL81****GAL4* (heterozygous at the *GAL81* locus) has a dominant constitutive phenotype (see previous sections). However, quantitative estimation of galactokinase indicated that ***GAL81*** *GAL4* is semi-dominant over its wild-type allele *GAL81* GAL4 in uninduced condition. Under

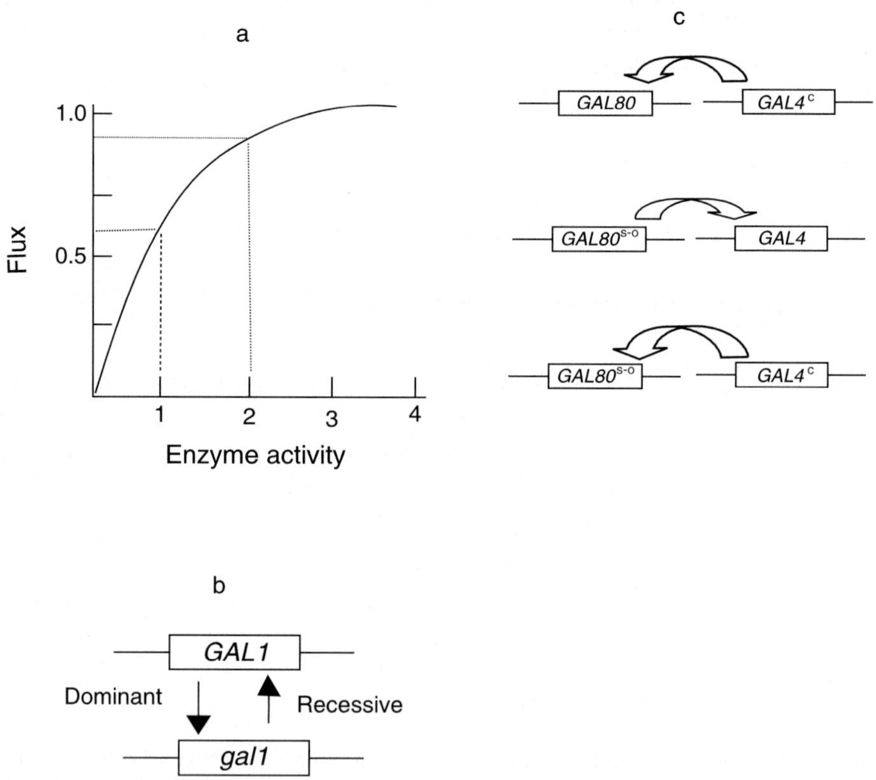

Fig. 4.3.1 Schematic representation of gene interactions. **a** Flux–enzyme activity relationship. Flux through a metabolic pathway is marginally reduced when the enzyme activity falls by 50% (from four units to two units) in a heterozygous diploid lacking one copy of the enzymes. If the wild-type diploid has low enzyme activity to start with, then a decrease in enzyme activity by 50% (say from two units to one unit) can reduce the flux significantly resulting in a dominant phenotype. **b** Relationship showing the dominance and recessiveness. **c** Epistasis

Table 4.3.1 Semidominance of *GAL4* constitutive allele over wild-type. *Bold* indicates the mutant allele (data obtained with permission from Matsumoto et al. 1980)

	Galactokinase activity	
Genotype	Uninduced	Induced
1. *GAL 81GAL4/GAL81GAL4* (Wild-type diploid)	0.01	5.27
2. ***GAL81****GAL4/****GAL81****GAL4* (Homodiploid)	4.16	5.07
3. ***GAL81****GAL4/GAL81GAL4* (Heterodiploid)	2.86	4.63
4. ***GAL81****GAL4/GAL81gal4* (Heterodiploid)	4.59	6.28

similar condition, ***GAL81*** *GAL4* was dominant over *GAL81gal4* (Table 4.3.1, row 3). This result is difficult to explain based on the classical operator-repressor model proposed by Douglas and Hawthorne. On the other hand, it is easily explained if one assumes that *GAL81* is a part of the coding region of *GAL4* rather than an operator locus. This is a classic example, wherein the genetic phenomenon can provide deep insights into the nature of interaction between the gene products. It turns out that detailed analysis eventually showed that in fact *GAL81* is a part of GAL4 protein itself.

4.3.3 Negative Dominance

If the protein is a homodimer, then mutant polypeptide coded by the defective allele can interfere with the normal polypeptide, which results in the reduction of functional protein to a level below the threshold required to maintain normal function. This type of dominance is different from the classical dominance, and is referred to as "negative dominance". Negative dominance occurs only when proteins are made up of more than one subunit. A well-studied example of negative dominance is a mutation in tumor suppressor gene p53 of humans, which normally exists as a homotetramer. Homozygous individuals bearing mutations in both the copies p53 gene are susceptible for tumor formation. Heterozygous individuals bearing one mutant and a wild-type copy also are susceptible for tumor formation. It is now known that the mutant p53 polypeptide interacts with the wild-type polypeptide, resulting in a non-functional protein.

4.3.4 Epistasis

In epistasis, a phenotype conferred by one mutant gene is influenced by a mutation in another gene. As mentioned in the previous chapter, the inability of $gal3\delta^{-}$ to grow on galactose is overcome by a recessive mutation in *GAL80*. Establishing an epistatic relationship between genes tells us about the hierarchy of gene interaction. Epistasis can be either recessive or dominant.

Recessive epistasis: Strains lacking uridyl transferase or epimerase do not grow on galactose as the sole carbon source but normally grow on glycerol or ethanol.

However, these strains cannot use glycerol or ethanol if galactose is present in the medium. This is because galactose-1-phospahte accumulates in strains lacking a transferase or epimerase in the presence of galactose. If such a strain is allowed to grow on a medium containing galactose plus ethanol, mutants come up that can grow on ethanol even in presence of galactose. Analysis of these mutants indicated that they contain mutations either in *gal1* or *gal4* locus. Accordingly, galactose is not toxic in a *gal1gal7* or *gal1gal10* double mutant, since galactose-1-phosphate is not formed. Loss of galactokinase function overcomes the toxic phenotype conferred by the defective transferase or epimerase. That is, a recessive mutation in galactokinase is epistatic over the *gal7* phenotype, indicating that it is a recessive epistasis. If mutations in two genes show an epistatic relationship, it is inferred that the two gene products interact at a physical or functional level. In this case, galactokinase and transferase interact at the functional level.

Dominant Epistasis: Galactose toxicity exhibited by *gal7* strains can also be suppressed due to dominant mutations. In fact, this trick was used by Douglas and coworkers to isolate dominant mutations at *GAL80* locus as discussed in the previous chapter. In this case, the dominance is due to the insensitivity of Gal80p to the inducer but retains the ability to keep the wild-type *GAL4* repressed. Therefore the mutation has caused only a partial structural alteration in Gal80p, suggesting that it is made up of independent domains.

Let us consider another example of epistasis involving *GAL4* and *GAL80*. A haploid strain bearing a dominant $GAL80^{s\text{-}0}$ allele renders the cell non-inducible. While this allele is dominant over its wild-type allele, it is epistatic over wild-type *GAL4*. In other words, $GAL80^{s\text{-}0}$ allele in some way prevents wild-type *GAL4* from functioning, even in the presence of the inducer. Similarly, ***GAL81*** allele is epistatic over wild-type *GAL80*.

According to the operator repressor model, ***GAL81*** represents a mutation in the operator of *GAL4* that renders normal Gal80p incapable of binding thus Gal4p is synthesized constitutively. On the other hand, $GAL80^{s\text{-}0}$ does not recognize the inducer, thus represses *GAL* expression even in the presence of the inducer. Thus $GAL80^{s\text{-}0}$ is epistatic over *GAL81* and ***GAL81*** is epistatic over *GAL80*. However, a haploid strain of the constitution ***GAL81*** $GAL80^{s\text{-}0}$ is constitutive, which means that ***GAL81*** is epistatic over $GAL80^{s\text{-}0}$.

The epistatic relationship between different alleles of *GAL 81* and *GAL80* took a curious turn when Oshima co-workers isolated $GAL80^{s\text{-}1}$ and $GAL80^{s\text{-}2}$, additional alleles of *GAL80*, as well as *GAL81-1* one more allele of *GAL81* locus. As shown in Table 4.3.2, the epistatic relationship depends on which alleles of *GAL80* and *GAL81* are under consideration. Based on these observations, Oshima suggested that the *GAL81* locus may not be an operator locus of *GAL4*, but might represent a special site on Gal4p through which Gal80p controls the activity of Gal4p. This idea in its simplest form suggests that most likely the Gal80p interacts directly with Gal4p.

Table 4.3.2 Induction status of the *GAL* system in haploid strains bearing varying combination of *GAL4, GAL81*, and *GAL80*^{s} alleles

Nature of alleles			Induction status	
GAL4	Operator	*GAL80*	Constitutive	Inducible
GAL4	*GAL 81* (WT)	*GAL80 (WT)*	No	Yes
GAL4	***GAL81*** (mutant)	*GAL80*	Yes	-
GAL4	***GAL81*** (mutant)	*GAL80*$^{s-0}$	Yes	
GAL4	*GAL81-1*(mutant)	*GAL80*$^{s-1}$	No	Inducible
GAL4	*GAL81-1*(mutant)	*GAL80*$^{s-2}$	No	No

Epistasis and genetic disorders: The above examples give a restricted view of the phenomenon of epistasis, but epistasis is a fundamental mechanism that plays an important role in modifying phenotypic expression at a much larger scale. We discussed classic galactosemia, a recessive monogenic disorder. It is observed that the severity of the classic galactosemia varies between different individuals bearing the same mutant allele, meaning that the variation in severity of the phenotype is not due to allelic variation. That is, the same mutation may cause less severe effects in one as compared to the other individual. Why is this so? This is because the individuals vary genetically at other loci. These genetic variations modify the phenotype through epistasis. Although galactose toxicity is widespread from bacteria to humans, a mouse defective for uridyl transferase accumulates as much galactose-1-phosphate as humans but do not exhibit any of the toxic phenotypes. This unusual deviation from expected phenotype could be a consequence of evolution of novel epistatic mechanism to overcome the toxicity.

4.3.5 Allele-Specific Interactions

The phenomenon wherein the phenotype varies depending upon the alleles under consideration is referred to as allele-specific interaction. As discussed in the previous chapters, intragenic complementation occurs only between specific alleles. Allele-specific interaction indicates a direct physical interaction between protein products. Allele-specific interactions can be understood by considering the lock-and-key analogy. According to the operator repressor model of Douglas and Hawthorne, Gal80p and the operator of *GAL4* (*GAL81*) bear a lock-and-key relationship while according to Oshimas model, Gal80p and Gal4p bear the lock-and-key relationship. In a lock-and-key analogy, a defective lock cannot be unlocked by a normal key, but the defect can be overcome by a complementary change in the key and vice versa. This is only possible if intermediary component is not required for the key to unlock. Similarly, strict structural complimentarity between the physically interacting proteins is the basis for allele-specific interaction.

References

Agius L, Aiston S, Newgard CB (2000) The control strength of glucokinase in hepatocytes: a predictor of metabolic defects in maturity onset diabetes of the young, Type 2. In: Cornish-Bowden AJ, Cardenas ML(eds) Technological and medical implications of metabolic control analysis. Kluwer Academic Publishers, Dordrecht, pp 109–115

Herskowitz I (1987) Functional inactivation of genes by dominant negative mutations. Nature 329:219–222

Kacser H, Burns JA (1981) The molecular basis of dominance. Genetics 97:639–666

Keightley PD (1996) A metabolic basis for dominance and recessivity. Genetics 143:621–625

Leslie ND, Lager KI, McNamara PD, Segal S (1996) A mouse model of galactose-1-phosphate uridyl transferase. Biochem Mol Med 59:7–12

Manson MD (2000) Allele-specific suppression as a tool to study protein–protein interaction in bacteria. Methods 20:18–34

Matsumoto K, Adachi Y, Toh-e A, Oshima Y (1980) Function of positive regulatory gene gal4 in the synthesis of galactose pathway enzymes in *Saccharomyces cerevisiae*. Evidence that the *GAL81* region codes for a part of the GAL4 protein. J Bacteriol 141:508–527

Nogi Y, Matsumoto K, Toh-e A, Oshima Y (1977) Interaction of superepresible and dominant constitutive mutations for the synthesis of galactose pathway enzymes in *S. cerevisiae*. Mol Genet 152:137

Stearns T, Botstein D (1988) Unlinked non-complementation: isolation of new conditional lethal mutations in each of the tubulin genes of *Saccharomyces cerevisiae*. Genetics 119:249–260

Wilkie AOM (1994) The molecular basis of genetic dominance. J Med Gen 31:89–98

4.4 Conditional Lethal Mutations

4.4.1 Introduction

In general, a mutation that inhibits growth under one set of conditions but not the other is called conditional lethal mutation. For example, a recessive mutation in *GAL1* results in conditional lethality, because it cannot grow on galactose as the sole carbon source, but can grow on glucose. That is, for a *gal1* mutant, a medium containing glucose is a permissive condition while galactose is non-permissive. We have discussed how to isolate conditional mutations in *GAL* gene(s), which are not essential for viability. How do we recover a strain bearing a mutation in an essential gene such as DNA polymerase, for example? Based on the property of proteins to function optimally in a narrow range of temperature, Horowitz and Leupold developed a method to isolate temperature-sensitive mutations in essential genes. Temperature-sensitive mutations provide a convenient handle to address questions that are otherwise not amenable for genetic analysis. For example, DNA polymerase of yeast functions optimally at 30 °C. It is possible to alter its structure by introducing amino-acid substitutions in a way that its activity is abolished at 35 °C but not at 30 °C. In general, only missense but not truncation or deletion mutants exhibit temperature-sensitive phenotype. The advantage of temperature-sensitive mutations is that the activity of the mutant protein can be reversed by changing the growth temperature.

4.4.2 Temperature-Sensitive Allele of GAL3

We know from previous studies that *GAL3* function is required only for rapid initiation of the induction of *GAL* genes. This probably implies that the *GAL3* function may not be required for the maintenance of the induced state but is required only for initiation of induction. *GAL3* function is similar to a car battery that is required for starting the engine. To investigate this possibility, Nogi isolated a strain bearing $gal3^{ts}$ allele (Fig. 4.4.1). He observed that the strain bearing $gal3^{ts}$ allele can be induced at 25 °C (permissive) but not at 35 °C (non-permissive). However, a strain bearing $gal3^{ts}$ allele induced at 25 °C (permissive) continued to maintain the induced state even after shifting the temperature to the non-permissive temperature. That is, once induced, the *GAL3* function is no more required for the inducted state (Fig. 4.4.2a). On the other hand, a strain bearing the $gal3^{ts}$ allele and a *gal1* defect, induced at 25 °C (permissive) was unable to maintain the induced state after shifting the temperature to 36 °C. This suggested that for the maintenance of the induced state in a strain lacking GAL3 function, normal galactose metabolic pathway is essential. This result and the result discussed in the previous chapter indicate that a *gal3* strain lacking

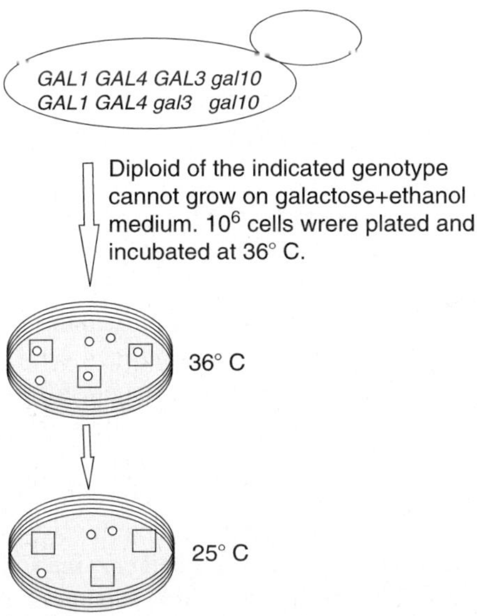

Fig. 4.4.1 Isolation of a temperature-sensitive allele of *GAL3*. For isolating the temperature-sensitive allele of *GAL3*, a diploid strain defective in both the copies of *GAL10* but defective in one copy of *GAL3* locus was used. Spontaneous mutants that cannot grow in ethanol plus galactose at 25 °C but can grow at 36 °C are sought from a diploid strain, which are likely to bear a temperature-sensitive allele of *GAL3*. The probability of obtaining t^s allele at *GAL1*, and *GAL4* loci is reduced by taking a strain heterozygous at *GAL3* locus, but not at *GAL1* and *GAL4*. The diploid mutant was sporulated to obtain a haploid with *gal3* t^s allele. This strain was mated to a wild-type haploid to outcross the recessive *GAL10* allele

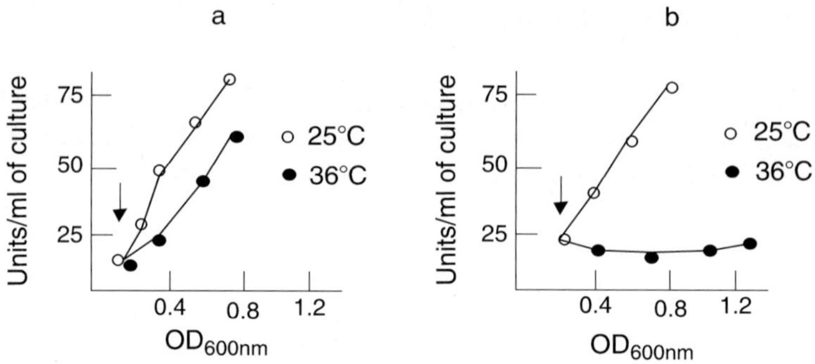

Fig. 4.4.2 *GAL3* function is required for the initiation but not for the maintenance of the induced state. **a** Haploid *gal3*^{ts} strain was grown in galactose and glycerol as the alternate carbon source up to an OD_{600} of 0.3. At this point, the culture was divided into two parts and the temperature of one of the culture was shifted to 36 °C (●) and the other was maintained at 25 °C (O). **b** A haploid strain of the constitution *gal1gal3*^{ts} was grown in glycerol and galactose at permissive temperature to an OD of 0.3 and then half of the culture was maintained at 25 °C and the other half at 36 °C. Uridyl transferase was monitored after the temperature shift in both experiments

either galactose metabolic pathway or mitochondrial function is unable to induce the *GAL* regulon.

4.4.3 Temperature-Sensitive Allele of GAL4

How does a recessive mutation in *GAL4* affect the expression of all the Leloir genes? Klar and Halverson reasoned that Gal4p functions as a common subunit of kinase, transferase, and epimerase. According to this idea, a recessive mutation in *gal4* would abolish the activity of all the Leloir enzymes. They tested this possibility by isolating *gal4*^{ts} allele using a similar strategy as described in Fig. 4.4.1. Leloir enzyme activity was measured in vitro at 35 °C in cell-free extract obtained from *gal4*^{ts} allele-bearing haploid strain grown in the presence of galactose at permissive temperature. It was no different from that of the enzymes activity present in cell-free extract obtained from a wild-type strain grown at the permissive temperature in presence of galactose. If Gal4p is a common subunit of Leloir enzymes, then the enzymes isolated form the strain bearing *gal4*^{ts} allele should not have exhibited enzyme activity in vitro at 35 °C. These results point out that Gal4p is not a common subunit of the Leloir enzymes as it was originally predicted.

Temperature-sensitive *GAL4* allele was used in an independent study to test whether Gal4p is synthesized constitutively as predicted by the operator repressor model. Matsumoto et al. reasoned that the induction kinetics of galactokinase expression in a strain bearing *gal4*^{ts} allele exposed to non-permissive temperature prior to the addition of galactose would be the same as compared to the normal strain in response to galactose at permissive temperature. This is because Gal4p is not synthesized prior to the addition of galactose as per the Douglas Hawthorne model. On the other hand,

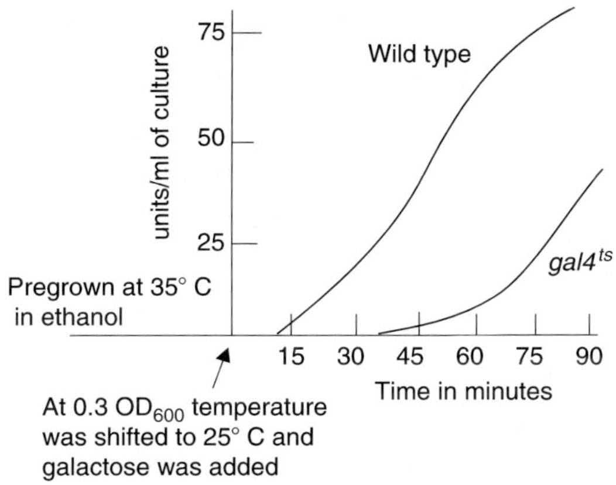

Fig. 4.4.3 Induction kinetics of galactokinase. Wild-type and mutant strain were grown in ethanol at 35 °C and when the OD_{600} reached 0.3, temperature was shifted to 25 °C and galactose was added. Galactokinase was monitored as a function of time

if Gal4p is synthesized constitutively, then exposure of the cells bearing *gal4^{ts}* allele to non-permissive temperature prior to the addition of galactose, would inactivate the existing Gal4p. This would delay the induction in response to galactose after shifting the cells to permissive temperature. In a strain bearing GAL4^{ts} allele grown at non-permissive temperature prior to the addition of galactose, galactokinase was expressed 35 min after galactose addition at permissive temperature (Fig. 4.4.3). Under a similar condition, the wild-type strain showed a lag of only 15 min. This result was interpreted to mean that Gal4p is synthesized even before galactose addition. Therefore, exposure of the mutant strain before addition of galactose had inactivated Gal4p, which existed before the addition of galactose. This is not what is expected if the operator repressor model were to be true.

The examples discussed above illustrate the utility of temperature-sensitive alleles in the analysis of gene functions. It should be born in mind, that in general, temperature-sensitive mutations are missense mutations that alter the temperature stability without drastically altering the inherent activity of the protein at the permissive temperature. Temperature-sensitive mutations do occur in nature and exhibit some interesting properties. For example, the dark extremities of the Himalayan cat are due to a temperature-sensitive mutation in a gene required for pigment formation. Because the body temperature is higher than the temperature at the extremities, this mutant allele is non-functional and the pigment formation does not occur, and hence the body appears non-pigmented while the extremities appear dark.

References

Edgar RS (1966) Conditional lethals. In: Cairns J, Stent GS, Watson JD (eds) The phage and the origins of molecular biology. Cold Spring Harbor Laboratory, Cold Spring Harbor, NY, pp 166–170

Nogi Y (1986) Gal3 gene product is required for maintenance of the induced state of the *GAL* cluster genes in *Saccharomyces cerevisiae*. J Bacteriol 165:101–106

Klar AJS, Halvorson HO (1974) Studies on the positive regulatory gene GAL4 in regulation of galactose catabolic enzymes in *Saccharomyces cerevisiae*. Mol Genet 125:203–212

Matsumoto K, Toh-e A, Oshima Y (1978) Genetic control of galactokinase synthesis in *Saccharomyces cerevisiae*: evidence for constitutive expression of the positive regulatory gene Gal4. J Bacteriol 134:446–457

4.5 Revised Model for *GAL* Genetic Switch

4.5.1 Introduction

Douglas and Hawthorne proposed that Gal80p blocks the expression of Gal4p in the absence of galactose by interacting with the *GAL81*, presumed operator locus of *GAL4*. In the presence of galactose, Gal80p is inactivated and is not capable of blocking the expression of Gal4p, which then activates the transcription of *GAL* genes. This implied that Gal80p does not interact directly with Gal4p and that Gal4p is synthesized only when galactose is present. Based on the nature of allele-specific interaction between *GAL4* and *GAL80* as well as the induction lag observed with *gal4*^{ts} allele, Oshima and coworkers hypothesized that Gal80p directly interacts with Gal4p to inhibit its transcription-activating function. This implied that Gal4p is synthesized regardless of whether galactose is present or absent. That is, Gal4p is synthesized constitutively, but its activity is inhibited by Gal80p in the absence of galactose. How do we test which one of the above models is correct?

4.5.2 Protein–Protein Interaction Model

Oshima and co-workers had a battery of strains each bearing a different non-inducible alleles of *GAL4*, of which *gal4.62* had a nonsense mutation located in the middle of *GAL4* ORF (discussed in section 3.1.5). Oshima reasoned that a constitutive or inducible revertant from a haploid strain of the genetic constitution *GAL80*$^{s-1}$*gal4.62* must be a rare event if the Douglas-Hawthorne model is correct (Fig. 4.5.1a). This is because, even if *GAL80*$^{s-1}$ gets mutated to a normal wild-type or a recessive allele, *GAL* genes will not be expressed since *gal4.62* is anyway non-functional. Neither would *gal4.62* reversion to wild-type *GAL4* convert to galactose-positive phenotype since the *GAL80*$^{s-1}$ allele would suppress the expression of Gal4p even in the presence of galactose. Therefore, a reversion of *GAL80*$^{s-1}$*gal4.62* strain to an inducible or constitutive one is possible only if two independent mutational events occur (Fig. 4.5.1a). That is, *gal4.62* should revert to wild-type *GAL4* and simultaneously *GAL80*$^{s-1}$ should get converted to *gal80* or *GAL80*. The frequency of these independent events is a product of the individual mutation frequency, and therefore is expected to be low.

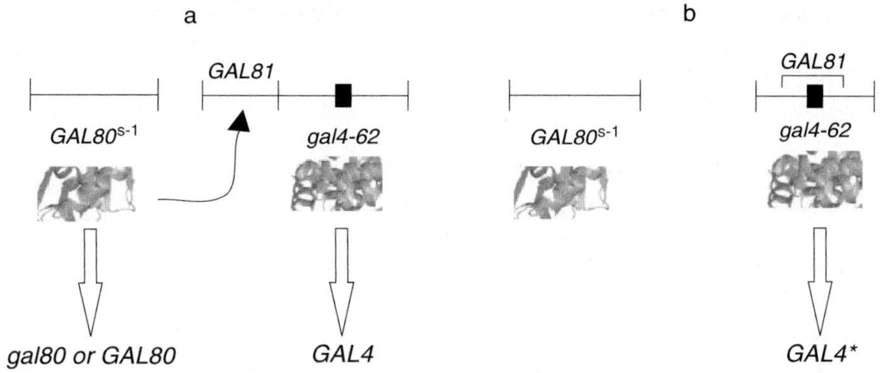

Fig. 4.5.1 Schematic illustration of possibilities of reversion to galactose-positive phenotype starting with a *GAL80*$^{s-1}$*gal4.62* strain. If possibility **a** is considered, two independent mutational events have to occur simultaneously to give rise to a galactose-positive phenotype. If possibility **b** is considered, mutation in *GAL4* locus alone can give rise to a galactose-positive phenotype, provided that *GAL4* allele thus generated (indicated as GAL4*) has to lose the ability to interact with *GAL80*$^{s-1}$ without losing the ability to activate the transcription. The second possibility was found to be true and the site of the constitutive mutation was mapped close to the site of nonsense mutation of *gal4.62* allele as indicated. That is, the *GAL81* region is apart of the Gal4p itself

On the other hand, if Gal80p and Gal4p proteins interact directly as Oshima proposed, it is possible to obtain a constitutive strain by just one mutational event. That is, if *gal4.62* allele mutates to a form capable of activating transcription but incapable of interacting with Gal80p, the strain would be constitutive. Oshima and co-workers isolated a constitutive strain starting from *GAL80*$^{s-1}$*gal4.62* mutant at a frequency expected of a single mutational event. On further genetic analysis, the secondary mutation responsible for the constitutive phenotype was found to be tightly linked to the nonsense mutation present in the *gal4.62* locus. This led to the suggestion that the original *GAL81* locus, which is thought to be a *cis*-acting element present adjacent to *GAL4* can no longer be considered as the operator site. Based on this, it was suggested that the repressor operator concept for *GAL* gene induction has to be revised.

4.5.3 Interaction Between GAL4 and GAL80 Proteins

Around the same time, Perlman and Hopper provided experimental evidence of different nature in support of the interaction between Gal4p and Gal80p. They used cyclohexamide, a eucaryotic translation inhibitor at a concentration which is known to inhibit the protein synthesis in yeast. They reasoned that the addition of cyclohexamide would effectively block Gal4p synthesis in response to galactose and therefore no transcription of *GAL1* or *7* would occur. However, they observed that galactose induced *GAL1* and *7* mRNA (the induction was monitored by in vitro translation) even if cells are exposed to cyclohexamide before the addition of galactose, at a concentration that effectively blocks in vivo protein synthesis. This clearly indicated that Gal4p was present even before the cells were exposed to cyclohexamide, suggesting that Gal4p is synthesized constitutively, a result that supports the Gal4p-Gal80p interaction and argues against the Douglas-Hawthorne model.

They also obtained evidence in favor of the protein–protein interaction model from synchronous mating experiments. Galactokinase activity cannot be detected in response to galactose in haploid strain of the genetic constitution (i) *gal1GAL4 GAL80* or (ii) *GAL1 gal4 GAL80* or (iii) *GAL1GAL4* $GAL80^{s-0}$. As expected, diploid formed between (i) and (ii) express galactokinase in response to galactose due to complementation as a function of diploidization (Table 4.5.1). However, when (i) was mated with (iii) in the continuous presence of galactose no galactokinase activity was detected (Fig. 4.5.2, Table 4.5.1). This is possible only if $Gal80^{s-1}p$ of the latter strain inactivated the Gal4p present in *gal1GAL4 GAL80* strain.

The above results clearly are not in conformity with the operator repressor model of Douglas and Hawthorne, but are compatible with that of the protein–protein interaction model. Although the evidences in support of the protein–protein model are circumstantial, the confidence in this model rests on the fact that the

Table 4.5.1 Results of the synchronous mating experiment. Galactokinase activity is expressed as nanomoles of galactose-1-phosphate formed per 10^8 zygotes (data obtained with permission from Perlman & Hopper 1979)

***a** GAL1gal4GAL80 X* **α** *gal1GAL4 GAL80*			***a** GAL11gal4*$GAL80^{s-0}$ *X* **α** *gall GAL4 GAL80*		
Time (h)	Kinase activity	Diploids (%)	Time (h)	Kinase activity	Diploids (%)
3	0	3.5	4	0	18
4	1.9	9.7	5	0	27
5	1.2	18	6	0	36

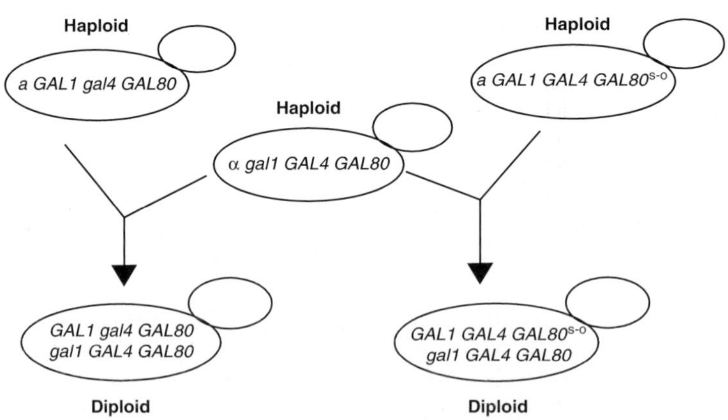

Fig. 4.5.2 Illustration of synchronous mating experiment. Haploid strains of indicated genotype were mated in the continuous presence of galactose. Galactokinase activity was measured as a function of diploidization

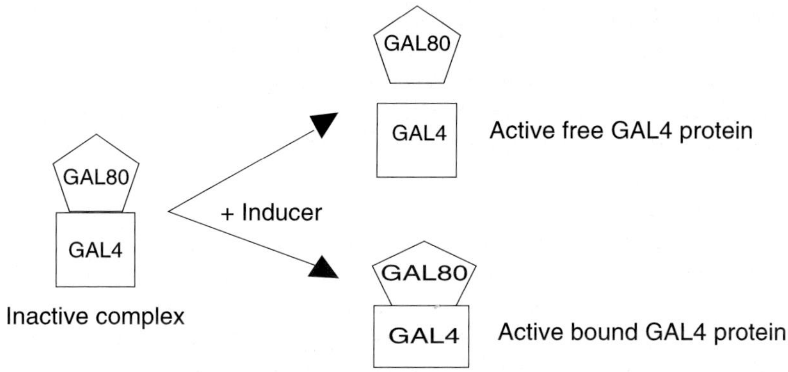

Fig. 4.5.3 Protein–protein interaction model. According to this, Gal4p and Gal80p physically interact and form an inactive complex in the absence of galactose. In the presence of galactose, Gal80p and Gal4p dissociate or reorient allowing Gal4p to activate transcription

rationale of each of these experiments is independent. The Gal4p-Gal80p interaction model depicted in Fig. 4.5.3 does not preclude the possibility of the inducer interacting with Gal4p.

References

Nogi Y, Matsumoto K, Toh-e A, Oshima Y (1977) Interaction of super-repressible and dominant constitutive mutations for the synthesis of galactose pathway enzymes in *Saccharomyces cerevisiae* Mol Genet 152:137–144

Perlman D, Hopper JE (1979) Constitutive synthesis of GAL4 protein, a galactose pathway regulator in *Saccharomyces cerevisiae*. Cell 16:89–95

Matsumoto K, Adachi Y, Toh-e A, Oshima Y (1980) Function of positive regulatory gene of Gal4 in the synthesis of galactose pathway enzymes in *Saccharomyces cerevisiae*: evidence that the *GAL81* region codes for part of the gal4 protein. J Bacteriol 141:508–527

Hopper JE (1981) Regulation of gene expression in the galactose mebiose regulation. In: von Wettstein D, Friis J, Kielland-Brandt M, Stenderup A (eds) Molecular genetics in yeast. Munksgaard, Copenhagen

Manson MD (2000) Allele-specific suppression as a tool to study protein–protein interaction in bacteria. Methods 20:18–34

4.6 Signal Transduction in *GAL* Regulon

4.6.1 Introduction

GAL3 was identified in 1948, more than a decade before *GAL4* and *GAL80* were identified. Progress in the understanding of *GAL3* function was marred by its unusual phenotype. Long-term adaptation was not in conformity with the conventional understanding of biological concepts and therefore approaches typically based on deductive logic were not applicable in deciphering the mechanism of Gal3p action. As a result, the focus got shifted to the analysis of *GAL4* and *GAL80* function in activating the expression of *GAL* genes. In fact, LTA has remained an enigma even after a fairly detailed understanding of the *GAL* genetic switch (discussed in later chapters).

As discussed before, during adaptation to galactose, *gal3* cells synthesize a factor which is lost by dilution during subsequent growth on carbon source other than galactose. Speigelman pointed out that the factor produced during adaptation is transmitted to its descendents through cytoplasm as long as galactose is present. This is an example of cytoplasmic inheritance. He also considered an alternate possibility that the inefficiency of *gal3* cells to adapt to galactose may be due to the presence of a factor that destroys the positive factor. The nature of the positive factor that comes into play in a *gal3* strain to establish the long-term adaptation nor the factor that is responsible for destroying the positive factor could not be deciphered from the available data.

4.6.2 Catalytic Model

Meanwhile, based on preliminary genetic data, it was hypothesized that Gal3p is an enzyme that synthesizes a coinducer. Based on circumstantial evidence, it was hypothesized that Gal3p is UDP glucose pyrophosphorylase, which catalyses the formation of UDPglucose from glucose1-phosphate and UTP. Further UDP glucose was presumed to be the inducer/coinducer that inactivates Gal80p, the repressor, to free Gal4p for activating transcription. One of the underlying assumptions of this model is that in a *gal3* strain there is a redundant mechanism to produce this inducer/coinducer albeit slowly, which is responsible for the long-term adaptation. However, this hypothesis did not stand the test of time.

After a decade, another hypotheses was advanced, which was similar to the one discussed above, except that it did not invoke the production of the coinducer. This is based on the observation that a *gal3* strain defective in any one of the Leloir genes, i.e., *gal3gal1, gal3gal7* or *gal3gal10* is not inducible by galactose. For example, *gal3gal1* strain cannot induce *GAL7* or *GAL10* in the presence of galactose. These results were interpreted in that Gal3p converts galactose to an inducer, which is a normal intermediate of galactose metabolism. Therefore, in the absence of *GAL3* and the normal galactose metabolic pathway, induction does not occur as the pathway for the inducer formation is absent. Accordingly, if only *GAL3* is absent, the normal galactose metabolite is made albeit at a slow rate through the normal galactose metabolic pathway. Thus, in a *gal3* strain, the eventual induction is due to the slow accumulation of these enzymes, which occurs in an autocatalytic way as expression of *GAL* enzymes otherwise requires an induction signal from galactose. We also learned from the previous chapter that the galactose metabolic pathway is also essential for the maintenance of the induced state in a *gal3* cell.

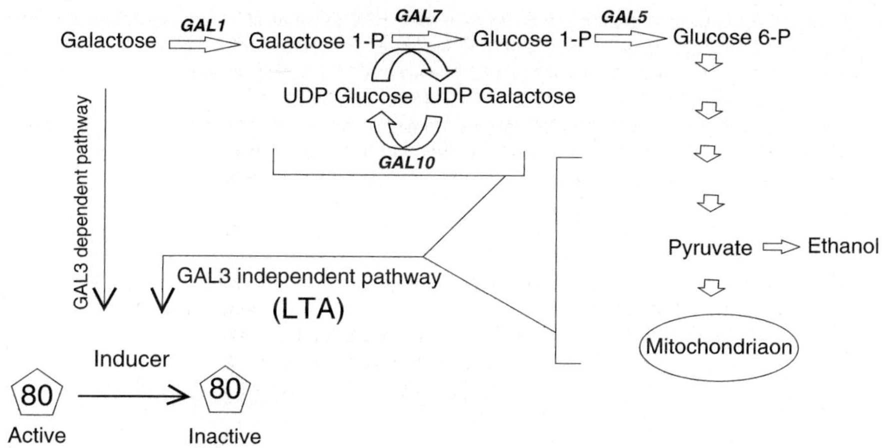

Fig. 4.6.1 Catalytic model. Gal3p converts galactose to a derivative that inactivates Gal80p the repressor. Inactivation of Gal80p leads to Gal4p-dependent transcriptional activation of *GAL* genes. In the absence of Gal3p-dependent signal-transduction pathway, galactose metabolism all the way through mitochondrial function is required for the manifestation of long-term adaptation (adapted with permission from Bhat et al. 1990)

Therefore, the normal galactose metabolic pathway seems to play an important role in the eventual induction and maintenance of the induced state in a *gal3* cell.

The above model predicts that mutations in enzymes not directly involved in galactose metabolism, for example, phosphoglucose isomerase, in conjunction with *gal3* mutation, i.e., *gal3pgi*, should exhibit long-term adaptation and not non-inducible phenotype. Surprisingly, the *gal3pgi* strain did not induce any of the *GAL* enzymes in response to galactose, suggesting that a galactose derivative formed by the activities of Leloir enzymes may not be the inducer. As discussed before, mitochondrial function was already known to be required for *gal3* strain to exhibit long-term adaptation. Based on these observations, it was proposed that a signal originating from galactose metabolism all the way through the mitochondrial function is needed for the *GAL3*-independent mechanism of induction, but the nature of this signal remained elusive for a long time (Fig. 4.6.1).

References

Bhat PJ, Oh D, Hopper JE (1990) Analysis of the GAL3 signal-transduction pathway activating GAL4 protein-dependent transcription in *Saccharomyces cerevisiae*. Genetics 125:281–291

Bhat PJ, Murthy TVS (2001) Transcriptional control of the GAL/MEL regulon of yeast *Saccharomyces cerevisiae*: mechanism of galactose mediated signal transduction. Mol Microbiol 40:1059–1066

Broach JR (1979) Galactose regulation in *Saccharomyces cerevisiae*. The enzymes encoded by the *GAL1, 7, 10* are coordinately controlled and separately translated. J Mol Biol 131:41–53

Tsuyumu S, Adams BG (1973) Population analysis of the deinduction kinetics of galactose long-term adaptation mutants of yeast. Proc Nat Acad Sci USA 70:919–923

Tsuyumu S, Adams BG (1974) Dilution kinetic studies of yeast populations: in vivo aggregation of galactose-utilizing enzymes and positive regulator molecules. Genetics 77:491–505

Chapter 5
Molecular Genetics of *GAL* Regulon

5.1 Cloning: A Perspective

5.1.1 Introduction

In general, a replica of a living or a non-living entity is called a clone. In biology, a clone refers to a progeny genetically identical to its parents. Clones are produced during vegetative or mitotic or asexual reproduction. Occasionally, during asexual reproduction, genetic variants arise at a frequency of 10^{-6} due to errors in copying the genetic information. Such variants are not the clone of the parent cell. In single-celled organisms like yeast, such variants can be isolated using genetic screens or selection, and this process is called "cloning of an organism". We have learned how to clone yeast strains that exhibit different phenotypes. Cloning multicellular organisms entails different approaches (Box 5.1.1). By definition, descendents of sexual reproduction are not clones because they are not genetically identical to the parents. This definition breaks down when one considers the haploid spores of a homo-diploid yeast cell. Here, the haploid spores produced through sexual reproduction are genetically identical to one another.

Just like we need clones of organisms for experimental analysis, we also need "clones" of proteins or DNA fragments for studying their physicochemical and biological properties. For example, a yeast cell can "clone" 10^6 galactokinase molecules upon induction by galactose and therefore it can be easily purified in large amount for further studies. On the other hand, in a haploid cell, only one copy of galactokinase gene exists and that too in covalent linkage with other genes. To obtain large amount of galactokinase gene, it needs to be separated from its genomic context and allowed to multiply in an appropriate host. The technique of isolating a gene from its natural context to make more of its copies is called "gene cloning". In principle, one can clone any piece of DNA, regardless of whether it codes for a specific cellular function or not. A battery of techniques collectively called recombinant DNA technology originated from independent studies of bacterial and phage genetics. Identification of plasmids and phages, the phenomenon of restriction modification and the transfer of plasmids or phages into the bacterium played a decisive role in the development of this technology, which revolutionized

P.J. Bhat, *Galactose Regulon of Yeast.*

Box 5.1.1 *Cloning of multicellular organisms*

Somatic cells of a multicellular organism such as human are produced by asexual or mitotic reproduction. Human contains clonal population of somatic cells derived from a zygote that is genetically unique. Because of this, the human individual is not a clone. That is, descendents of sexually reproducing organisms are not clones. It is now possible to clone sexually reproducing multicellular diploid organisms such as sheep. For example, Dolly was cloned by transferring the nucleus containing the genetic material taken from the cells of udder of a 6-year-old ewe into an ovum that was denucleated. This zygote was implanted in a surrogate mother and allowed to develop into a full-fledged sheep, Dolly. Therefore, Dolly is a clone since she is genetically identical to the ewe from which the genetic material was taken. In principle, it was possible to generate clones of Dolly using the genetic material from the udder of the same ewe. For a long time it was believed that the genetic material undergoes irreversible changes during differentiation that it cannot be turned back to initiate a fresh developmental program, which proved to be incorrect with the cloning of Dolly. Very rarely, in multicellular organisms, the first two cells of zygotic division develop into two genetically identical individuals. Such individuals are referred to as fraternal twins who are clones of one another.

the study of biology beyond comprehension. In the ensuing section, I have briefly discussed its origin and how quickly it was inducted to unravel yeast biology.

5.1.2 Vectors, Genetic Transformation, and Recombinant DNA Technology

The first plasmid to be discovered was the F (fertility) plasmid of *E. coli*, which transferred genes from one bacterial chromosome to another through a process known as conjugation. Bacterial strains resistant to many antibiotics were first observed in Japan through the analysis of an epidemic of bacillus dysentery. This antibiotic resistance was found to arise by conjugational transfer of genes that code for antibiotic resistance factors present on plasmids from certain antibiotic resistant *E. coli*. Certain bacteria also carry plasmids that code bacteriocinogens that kill closely related bacteria not carrying the plasmids. In general, plasmid-borne traits are not essential for the organism, but they encode special features, which may provide advantages to the cell under special circumstances. Plasmids are circular DNA molecules that replicate independently of chromosomes and can move from cell to cell depending upon whether they have the genes for conjugational transfer. This property of the plasmids has been exploited to use it as a vehicle or vector to carry genes.

In the early 1970s, it was demonstrated that antibiotic-sensitive bacterial cells treated with calcium chloride become competent to take up the purified plasmid bearing the antibiotic resistance genes. The *E. coli* cells that have picked up the drug resistance factors could easily be identified by plating the cells on a medium containing the drug. These observations raised the possibility that in vitro-constructed DNA molecules can also be introduced into bacteria, provided an experimental technique to identify the bacteria that have taken up the DNA exist. These naturally occurring plasmids have since been engineered by incorporating features that allows the ligation of any DNA and its subsequent identification. This was the beginning of the recombinant DNA technology and the reverse genetic approach that revolutionized biology.

Bacteria can defend against infecting phages by attacking the intruding DNA at specific sites. This primitive immune system of bacteria is called "restriction" and the specific sites where the DNA gets cut is called the "restriction site". A nucleolytic enzyme capable of cutting the invading phage DNA in this manner is called a restriction endonuclease. How is the host DNA protected from its own restriction endonuclease? This is achieved by the modification of the restriction site by methylation of certain bases within the vulnerable sites such that the host restriction enzyme does not cut its own DNA. If a bacteria carries a restriction enzyme that recognizes say a 6-bp sequence, it has to have a modification system to protect the site from self-destruction. This fundamental observation led to the identification of a large number of restriction enzymes bearing different specificities from different bacteria.

Despite the restriction modification systems, many phages were identified that invaded and integrated into the bacterial genome (prophage). A noteworthy example of this is the λ phage that exists as a linear phage with cohesive ends. After the phage injects the linear DNA, it circularizes and the cohesive ends are ligated by the ligase of the *E. coli* to make the linear phage DNA into a covalently closed circle. This molecule then integrates into the circular bacterial genome and propagates as a part of the bacterial chromosome. For some reason, if the host is unable to propagate, the phage DNA is excised, multiplied many times, phage particles are assembled and released from the host to infect fresh neighboring *E. coli*. Since λ phage can enter an *E. coli* and multiply within the cells just like the plasmid, it also became a favorite vector for carrying DNA fragments.

The observation that the cohesive ends of λ phage are ligated in vivo indicated that any two DNA molecules can be ligated in vitro, provided cohesive ends can be

Table 5.1.1 List of some naturally occurring plasmids

Plasmids	Organism	Copy number	Size (in kb)	Function
ColE1	*E. coli*	30	9	Colicin E1 production
F	*E. coli*	2 to 3	95	Sex factor
R6K	*E. coli*	40	40	Antibiotic resistance
2 μ	*S. cerevisiae*	30	95	No known function

generated. Paul Berg's laboratory for the first time demonstrated the success of this technique. He constructed a hybrid genome (a recombinant molecule) of a 10-kb pair plasmid obtained form *E. coli* and a 5,432-kb pair a SV40 viral genome. This involved linearizing the two circular DNA molecules using a restriction enzyme, generating cohesive ends and ligating the DNA molecules with ligase in vitro. Since then, construction of hybrid DNA molecules or recombinant molecules and genetic transformation of many organisms has become routine.

5.1.3 DNA Cloning

The first step in cloning a gene is to obtain all the genes of an individual organism as separate molecular entities. That is, the genes are no more physically contiguous with each other as they exist in the genome (Fig. 5.1.1). In the next step, the genes are ligated to a vector such as a plasmids or phages or artificial chromosomes using recombinant DNA techniques and are introduced into the appropriate host such as *E. coli*. During this protocol, each *E. coli* cell takes up individual recombinant mol-

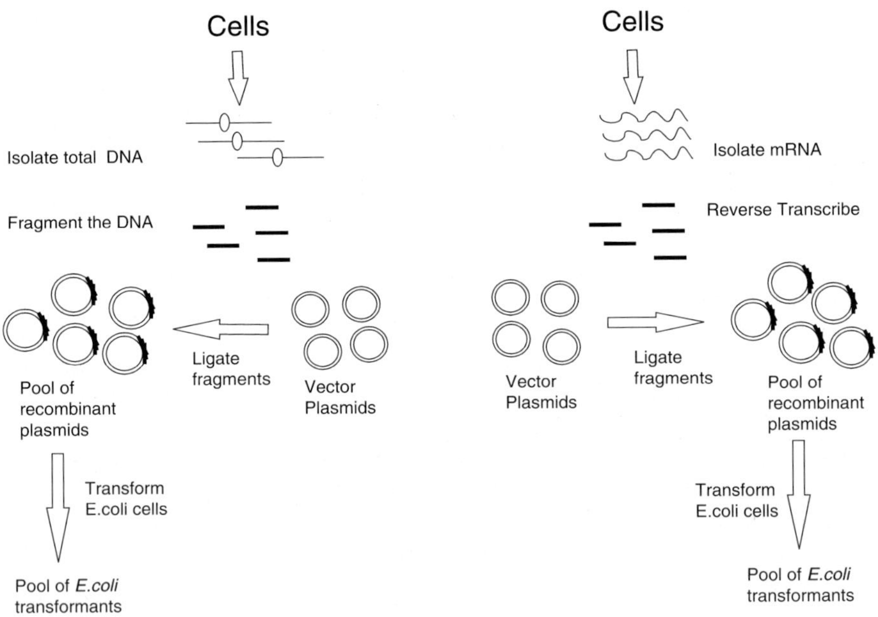

Fig. 5.1.1 Construction of gene libraries. Fragments obtained form the genomic DNA or from mRNA by reverse transcription and second-strand synthesis are ligated to vectors at convenient restriction sites. The pool of in vitro constructed recombinant molecules are transferred to *E. coli* by transformation. *E. coli* harboring the recombinant molecules are pooled and stored for future use. Whenever the library of recombinant molecules is required, *E. coli* are allowed to grow in selective medium and the recombinants are extracted free of the chromosomal DNA for further use. Alternately, *E. coli* bearing the desired recombinant is isolated from the pool, using specific techniques

ecules. The host multiplies the recombinant molecule to large numbers and they can be purified independent of the host chromosomal DNA. Such a population of recombinant molecules consisting of different genes but the same vector backbone serves as a library of genes from which specific genes can subsequently be isolated (Fig. 5.1.1).

Thus, a library of genes say from humans or any organism of one's choice can be constructed, stored, and propagated whenever required. The gene libraries are mainly of two types: genomic and cDNA, based on how the DNA fragments are obtained. The choice of the type of library for subsequent isolation of a specific gene depends on many factors.

The accelerated pace of developments in technology has resulted in newer approaches for cloning genes. Currently, PCR-based technology is widely used for cloning genes directly from the genome, circumventing the need to construct gene libraries. Nevertheless, library-based cloning is still in use and serves many other purposes. For example, a library serves as a constant source of genes of an organism without going back to the organism whenever the genomic DNA is required.

5.1.4 Genomic DNA Library

Genomic DNA is digested into random fragments of desired length with restriction enzymes. The length of the fragments to be used for constructing genomic library is based on the purpose for which the library will be used. If the library is to be used for genome mapping, then it is desirable to have larger DNA fragments, but as the fragment size increases, the choice of the vectors becomes restricted. For example, large fragments of say 50 kb cannot be cloned into plasmids but can be cloned into phages or artificial chromosomes. On the other hand, if the library is used for isolating protein coding sequences, then the average length of DNA fragments would depend on the average size of the genes and its flanking sequences. For example, for yeast, 1,700 clones of 35-kb fragment would represent the whole genome and one can find a gene of ones interest from this library with a probability of 0.99. It is possible to calculate the number of clones that must be present in a library to give a probability of obtaining a particular DNA sequence. $P = 1-(1-f)^N$ where f is the size of the fragment expressed as a fraction of the genome, P is the probability and N is the number of recombinants.

5.1.5 cDNA Library

Total mRNA is converted to double-stranded molecules using standard techniques and ligated to vectors. cDNA libraries enriched for recombinant clones of one's interest can be constructed provided prior information on the nature of the gene to be isolated is available. For example, separate cDNA libraries can be constructed from mRNA obtained after size fractionation. mRNA obtained from specific cell types or from cells grown under different experimental conditions can also yield cDNA library enriched for clones of one's interest. Recombinant clones of cDNA library do not contain the sequence elements normally associated with the gene such as introns. The vector for constructing cDNA library contains a promoter and a terminator in addition to other features. Since promoters are species-specific, the vectors will have to be designed depending upon the host in which the library is going to be used.

5.1.6 Isolation of Recombinant Clones

Convincing evidence for a genetic role of DNA was obtained by demonstrating that a genomic fragment could restore the lost function to a mutant cell. This demonstration was possible only after methods to make the cell competent to take up externally added DNA became available. This approach and the fact that a clonal population of micro-organism can be recovered from cells transformed with DNA made it possible to screen or select transformants containing any desired gene provided appropriate strategy to identify such transformant is available. Diverse techniques and approaches have been developed for isolating an *E. coli* transformant bearing the recombinant molecule of one's choice (see below). These fall mainly into three major categories.

The individual clones can be identified based on (a) hybridization using DNA probes, provided prior knowledge of the DNA sequence of the genes to be isolated is available, such as protein sequence from which the DNA sequence can be derived, or based on the sequence of the homologues (b) probing the protein product formed by the transformant using antibodies or any interacting molecule. For example, an *E. coli* transformant bearing a recombinant plasmid coding for a DNA-binding protein can be screened by using a labeled target DNA fragment. (c) the function of the expressed protein. Here in vivo screening techniques such as complementation or suppression of a given phenotype can be used. These techniques are generally used in microorganisms such as *E. coli* or yeast. In fact, many human genes have been isolated by their ability to functionally complement the corresponding function or to suppress some phenotypes in *E. coli* or yeast (Box 5.1.2). Many variations of the above approaches can be developed on a case-by-case basis.

Isolation of a gene using PCR: This technique has had a major impact in molecular biology. It is used to amplify a precise fragment of DNA from a complex mixture of starting template DNA. The only information required is the sequence of the primers that flank the DNA to be amplified. For example, if we know the amino-acid sequence of a protein, one can design degenerate primers (because of the codon redundancy) to amplify the corresponding gene directly from the genomic DNA or from libraries. PCR based cloning strategy has replaced in many cases the traditional cloning strategy.

Box 5.1.2 *Human and yeast galactokinase genes complement a galK⁻ E. coli for growth on galactose*

The amino-acid sequence of one of the tryptic peptides of human galactokinase matched with the sequence of a cDNA clone of T-cell cDNA library. This clone contained a cDNA capable of encoding a protein of 392 amino acids. A *gal*K$^-$ *E. coli* transformed with a expression plasmid bearing the cDNA grew on galactose as the sole carbon source. This result was taken to indicate that the plasmid indeed codes for human galactokinase which was shown to be true by further experimentation. Functional complementation was also used to isolate yeast galactokinase gene. Here, yeast recombinant clone was isolated from the genomic library by identifying the galK$^-$ *E. coli* transformant that is capable of growing in galactose as the sole carbon source. In fact many human genes have been isolated by functional complementation in yeast.

5.1.7 Development of Yeast Shuttle Vectors

The success of isolating a yeast gene came rather fortuitously. Yeast genomic DNA was fragmented, ligated at the *Eco*RI site of the ColEI (see Table 5.1.1). This mixture of recombinant molecules was introduced into *E. coli* mutant unable to grow in the absence of exogenously added leucine due to a specific defect in *leuB* gene encoding β isopropylmalate dehydrogenase. Surprisingly, one of the hybrid plasmid pYeleu10 containing the functional counterpart of the *E. coli leu*B gene complemented the genetic defect of *E. coli*. Since this plasmid attains a copy number of 30 per cell in *E. coli*, it could be easily purified, and its ability to transform a *leu2* (this gene is the functional counterpart of *leu*B of *E. coli*) yeast strain was tested. Only, few yeast transformants capable of growing on a medium lacking leucine could be isolated, indicating that the transformation efficiency was low. Further analysis of the *LEU2* transformant indicated that the whole plasmid had integrated into the yeast genome at the *LEU2* locus and the plasmid did not exist as an episome (Fig. 5.1.3). Since the integrating plasmids are difficult to retrieve, efforts to develop plasmids that can replicate in yeast as well as *E. coli* were developed. Nevertheless, the observation that *LEU2* gene integrates at its cognate site in the genome due to homologous recombination was a fundamental observation that propelled yeast molecular genetics to a different level.

The first autonomously replicating yeast vector plasmid was a derivative of pYeleu10 bearing a segment containing the origin of replication of the endogenous 2 μ plasmid of yeast (see Table 5.1.1). This hybrid plasmid transformed a yeast *leu2* strain to leucine prototrophy at high frequency without integration into the chromosome. Moreover, this plasmid could be isolated from the yeast cells and amplified in *E. coli* and retransformed back into yeast, indicating that the

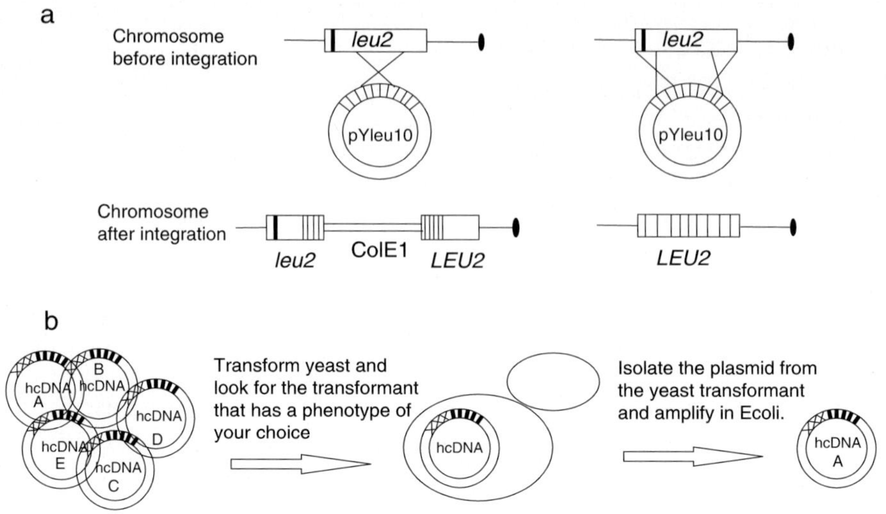

Fig. 5.1.2 Genetic transformation of yeast. **a** The defective *leu2* of the recipient yeast strain is indicated by a *box with a vertical bar*. The striped area of pYeleu10 represents the wild-type genomic fragment corresponding to the *LEU2* locus while the rest represents the ColE1 sequences. Due to a single crossover, the whole plasmid integrates at *LEU2* locus. *LEU2* locus is duplicated, of which one is a mutant and the other is a wild-type (*left panel*). If a double cross event occurs, the genomic *leu2* locus is substituted by the wild-type *LEU2* locus (*right panel*). The structure of the chromosome after the integration was deduced by Southern blot hybridization. **b** Illustration of isolating a human cDNA recombinant clone from a library of human cDNA plasmids by functional complementation. Transform a yeast strain with the hcDNA library, (prepared in a replicative yeast –*E. coli* shuttle expression plasmid wherein the cDNA expression is driven by a yeast promoter) and select yeast transformants based on the marker present on the plasmid. Of these, screen the transformants expressing the desired phenotype, extract the hcDNA clone and retransform *E. coli* for amplification and further characterization

Box 5.1.3 *Identification of yeast chromosomal elements*

As mentioned in the text, recombinant plasmids containing yeast *ARG4* and *TRP1* were also isolated from the yeast genomic library constructed in ColE1, by complementation of *E. coli* mutant lacking the corresponding function. Unlike pYEleu10, the ColE1 plasmid containing the *ARG4* or *TRP1* transformed the corresponding yeast mutant with high frequency. Detailed analysis indicated that the yeast genomic fragment containing *ARG4* and *TRP1* genes were closely linked to autonomously replicating sequences (ARS for chromosomal autonomously replicating sequences), by virtue of which the recombinant plasmids transform yeast with high frequency and exist in episomal form. These and other observations led to the development of yeast and *E. coli* shuttle plasmids. Later on, yeast centromeric sequences isolated by positional cloning approach (see section 5.5.6) were used in CEN based plasmids.

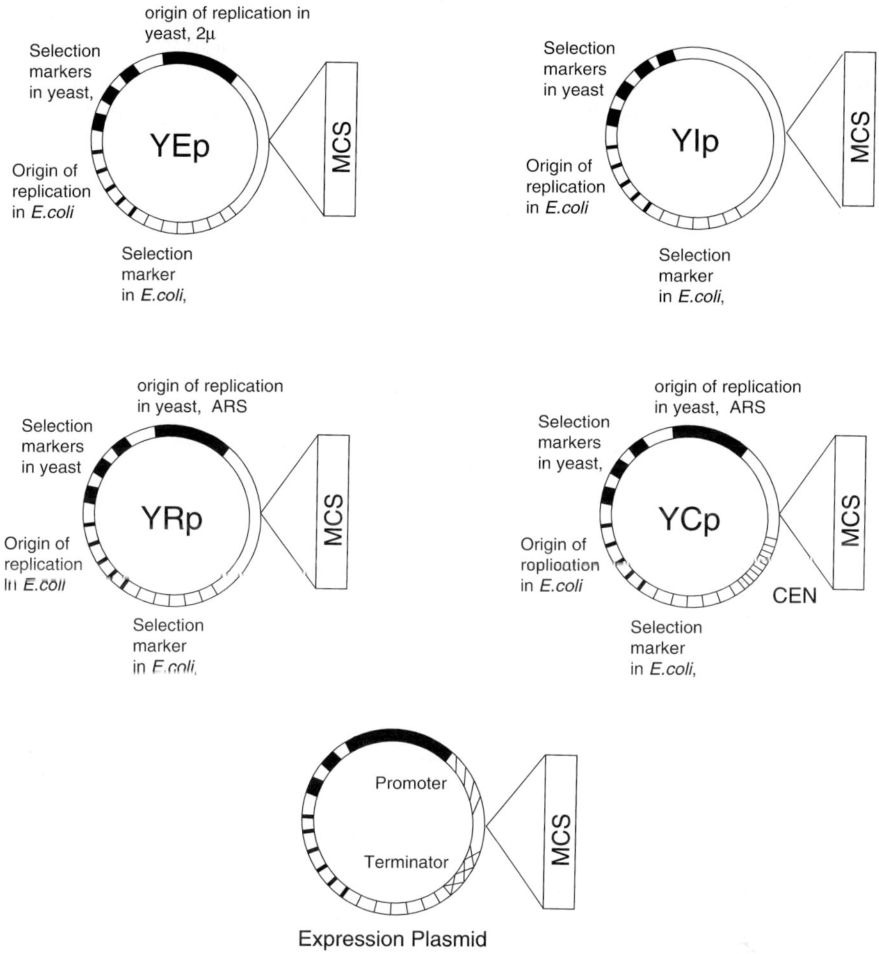

Fig. 5.1.3 Yeast *E. coli* shuttle plasmids. Plasmid maps of yeast *E. coli* shuttle plasmids (maps are not drawn to the scale). YEp, YCp,YIP and YRp have a bacterial origin of replication and lactamase as a selection marker. They have an auxotropic marker such as *URA3, HIS4, TRP1* or *LEU2* for selection in yeast. YEp has 2 μ plasmid origin of replication, while YRP has the chromosomal ARS (see Box 5.1.3) sequences. YCp is a derivative of YRp with centromeric sequence. This sequence provides mitotic and meiotic stability and is maintained roughly 1 copy per cell. YIp does not replicate in yeast as it does not have sequences required for autonomous replication. Most of the plasmids currently available have multicloning site. Expression plasmids are constructed by introducing yeast promoter and terminator sequences into YEp and YCp. YEp24 and YEp 13 were the first replicative shuttle plasmids constructed with *URA3* and *LEU2* auxotropic markers, respectively

plasmid had retained its structural and functional integrity. Based on this success, four types of yeast shuttle plasmids YEp, YRp, YCp and YIp were then constructed (see Fig. 5.1.2). These plasmids have been derivatized to include many convenient features for cloning and expression. For example, yeast expression plasmids are derivatives of YEp or YCp vectors with a yeast promoter followed by multicloning sites and a terminator. Expression of the ORF cloned at the multicloning sites would be driven by the features dictated by the promoter. Expression plasmids mainly serve two purposes. To characterize the *cis*-acting promoter elements as well as to express the proteins of one's interest. For example, a human cDNA library prepared in yeast expression vector can be used as a source for isolating human genes based on functional complementation (Box 5.1.2).

References

Adams A, Gattschling DE, Kaiser, CA, Stearns T (1998) Methods in yeast genetics. Cold Spring Harbor Laboratory Press, New York

Beggs JD (1978) Transformation of yeast by a replicating hybrid plasmid. Nature 275:104–109

Botstein D, Davis RW (1982) Principles and practice of recombinant DNA research with yeast. In: Strathern JN, Hicks JB (eds) The molecular biology of the yeast *Saccharomyces*: metabolism and gene expression. Cold Spring Harbor Laboratory, Cold Spring Harbor, New York, pp 607–636

Citron BA, Feiss M, Donelson JE (1979) expression of the yeast galactokinase gene in *E. coli*. Gene 6:251–264

Cohen SN, Chang ACY, Hsu L (1972) Nonchromosomal antibiotic resistance in Bacteria: genetic transformation of *E. coli* by R-factor DNA. Proc Nat Acad Sci USA 69:2110–2114

Carbon J (1993) Genes, replicators, and centromeres: the first artificial chromosomes. In: Hall MN, Linders P (eds) The early days of yeast genetics. Cold Spring Harbor Laboratory Press, New York, pp 375–390

Davis RH (2003) The microbial models of molecular biology. From genes to genomes. Oxford University Press

Hall BD (1993) Starting to probe for yeast genes. In: Hall MN, Linders P (eds) The early days of yeast genetics. Cold Spring Harbor Laboratory Press, New York, pp 391–404

Hicks JB, Hinnen A, Fink GR (1980) Properties of yeast transformation. Cold Spring Harbor Symp Quant Biol 43:1305–1313

Struhl K (1983) The new yeast genetics. Nature 304:391–397

Struhl K, Stinchcomb DT, Scherer S, Davis RW (1979) High-frequency transformation of yeast: autonomous replication of hybrid DNA molecules. Proc Nat Acad Sci USA 76:1035–1039

5.2 Genomic Organization of *GAL* Cluster

5.2.1 Introduction

Gal4p activates transcription of *GAL* genes probably by recognizing the 5′ regulatory sequences. Therefore, it is necessary to isolate these sequences to understand how Gal4p interacts with these sequences to activate transcription. Although a yeast genomic library would represent recombinants containing the 5′ regulatory regions, no molecular probe was available to isolate these sequences. These sequences can be isolated from the library by virtue of the fact that they are in physical contiguity with the coding sequences of *GAL* genes. How can one fish out the recombinants containing the *GAL* genes from the genomic library? A yeast genomic library constructed by ligating *Hin*DIII digested yeast genomic DNA to phage λ590 consisting of approximately 1.2×10^4 recombinants representing the whole yeast genome was screened for the *GAL* genes.

5.2.2 Cloning of the GAL Cluster

The sequence of *GAL* genes was not available to generate DNA probes for screening the desired recombinants from the library. This problem was circumvented by preparing a pool of radioactive cDNA probes from two sets of total mRNA. One set of mRNA was isolated from galactose grown cell and the other from glucose. The former contains labeled cDNA complementary to *GAL* genes in addition to other genes, but the latter is not expected to contain labeled complimentary sequences to *GAL* genes, because *GAL* genes are repressed in glucose grown cells. Therefore, recombinant phage population hybridized with ^{32}P labeled cDNA probe prepared from mRNA isolated from galactose and not glucose grown cells is expected to light up spots glucose grown cells (spot indicated by the arrow in Fig. 5.2.1b, was not lit up by cDNA probe obtained from glucose grown cells). Using this differential hybridization technique, a phage clone λgt1-Sc481 was isolated from the genomic library. The yeast genomic DNA fragment present in λgt1-Sc481 was subsequently used as a specific DNA probe to pull out recombinant phage clones that had overlapping DNA sequences. This exercise yielded three more clones λgt1-490, λgt1-491, and λgt1-494. A detailed restriction mapping of these clones and the genomic organization of the *GAL* cluster with its flanking sequence are shown is shown (Fig. 5.2.2).

Yeast genomic DNA fragments present in λgt1-481, λgt1-490, λgt1-491 and λgt1-494 recombinant phage clones represent gene(s) whose transcription is induced by galactose and not necessarily represent only *GAL* genes. To confirm that the above genomic fragments contain *GAL* genes, specific restriction fragments obtained from these clones were used as probes to detect the transcripts. Three distinct transcripts were detected from mRNA isolated from galactose but not glucose grown cells using different probes (Fig. 5.2.2). Based on the sizes of the mRNA and the known molecular weight of the corresponding proteins, it was inferred that the cloned DNA fragment consists of *GAL1, 7*, and *10*. Northern blot data was also consistent with the observation that these three genes are tightly linked and that *GAL* genes are separately transcribed from independent promoters.

The direction of transcription of the three genes was determined using a sandwich technique. λgt1-Sc481 DNA (which contains phage vector sequences as well as the *GAL* cluster) strands were separated and hybridized with GAL1, 7, and 10 mRNA on a nitrocellulose filter The mRNA:DNA hybrids were then detected using strand-specific ^{32}P labeled λ 590 DNA. GAL10 and GAL7 transcripts

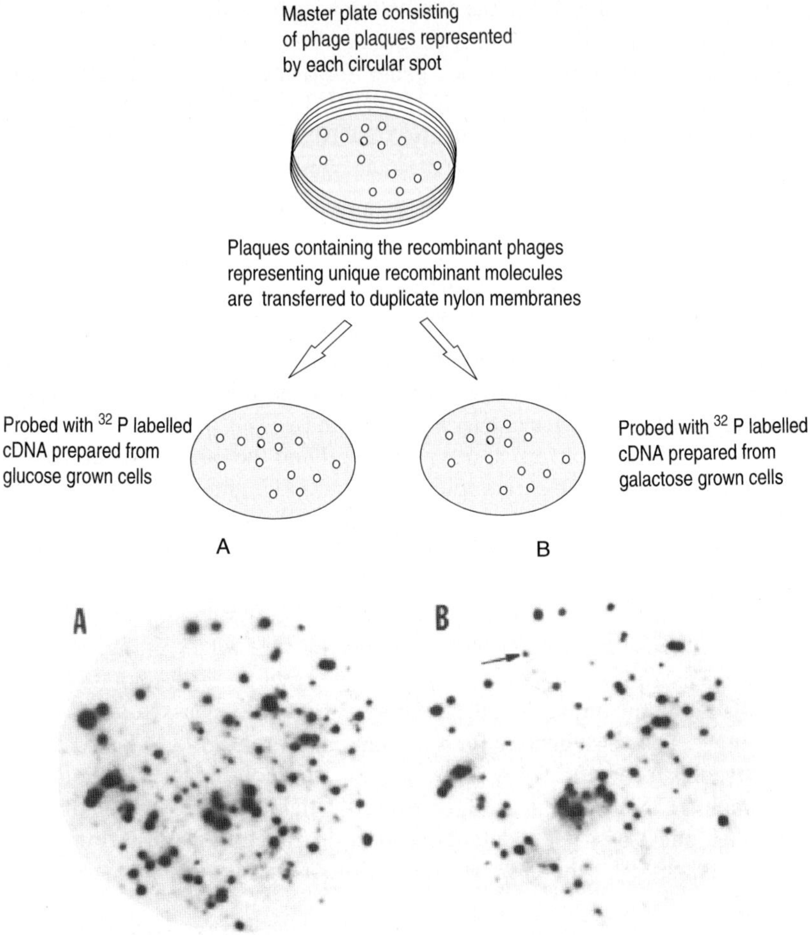

Fig. 5.2.1 Isolation of recombinant phages containing *GAL* gene cluster. Plaques containing the recombinant phage DNA are transferred to duplicate nitrocellulose filters that are probed as described. The recombinant plaque indicated by the *arrow* (filter B) was recovered from the master plate and reamplified and analyzed (adapted with permission from John and Davis 1979)

were picked up by the same strand of λ590 while GAL1 mRNA was picked up by the complementary strand. This result suggested that *GAL7* and *10* are transcribed in tandem from the same strand while *GAL1* is transcribed divergently from a complementary strand. The transcription initiation of *GAL1* and *10* was determined using both S1 nuclease protection assay as well as primer extension (Fig. 5.2.3). Based on the above analysis, it was inferred that the *GAL1-10* promoter should lie in restriction fragment E (see Fig. 5.2.2) spanned by *EcoR* 1 and *Ava*1 restriction site. This fragment was sub-cloned and sequenced (Fig. 5.2.3). Based on the available N-terminal amino-acid sequence of Gal1p, the position of

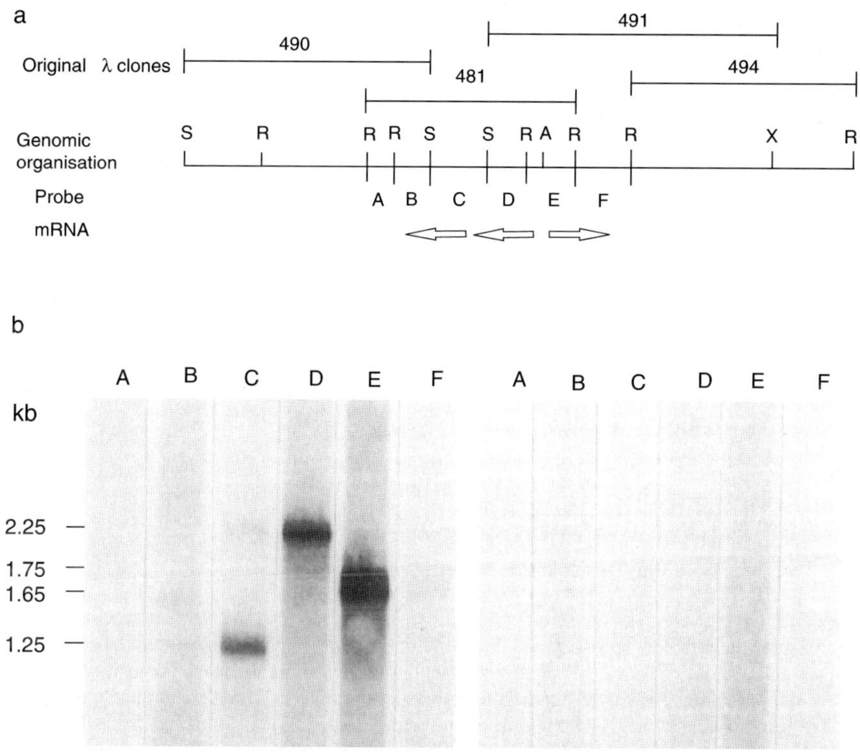

Fig. 5.2.2 Genomic organization and Northern blot analysis of *GAL* cluster. **a** λgt1-Sc481 is the initial recombinant clone isolated form the library. Subsequently, λgt1-Sc490, λgt1-Sc 491 and λgt1-Sc494 independent clones were isolated form the library. Based on the restriction pattern of these individual clones, genomic organization with relevant restriction enzyme sites was reconstructed (E, *Eco*RI, S, *Sal* I, A, *Ava* I X, *Xho*I). **b** Total mRNA isolated from galactose (*left panel*) or glucose grown (*right panel*) cells, was electrophoresed, transferred to nitrocellulose filters, and probed with different radioactively labeled restriction fragments. Each lane containing the total mRNA was separately probed (adapted with permission from John and Davis 1981)

start codon of galactokinase was identified (Fig. 5.2.4). There are two potential in frame ATGs for GAL10, one located at 33 to 31 and the other 139 to 137. Since the *GAL10* transcription initiation was at position 150, it was inferred that the likely ATG of *GAL10* is from 33 to 31.

5.2.3 Analysis of GAL1-10 Intergenic Region

Deletion analysis was used to define the *cis*-acting elements required for the galactose inducibility. In one approach, the 914 base pair intergenic region was cloned into a plasmid. β-galactosidase gene was fused with *GAL10* or *GAL1* translational frame

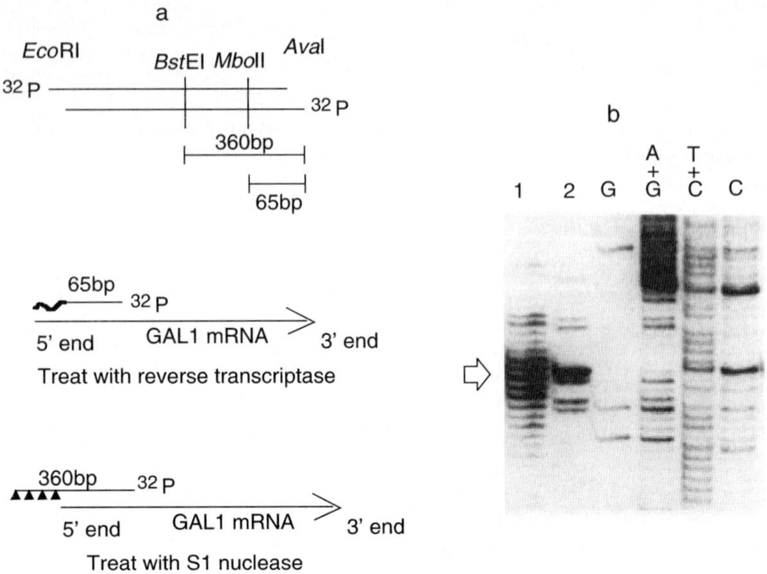

Fig. 5.2.3 Determination of transcription initiation of GAL1. **a** Total mRNA was separately hybridized in solution with 360-bp *Bst*E1 and Ava1 and 65-bp *Mbo*II and *Ava*I fragment labeled with ^{32}P at *Ava*1 end. The mRNA and 360-bp hybrid was digested with S1 nuclease (shown by *arrows*) to cleave the overhang. The mRNA and 36-bp hybrid were treated with reverse transcriptase to synthesize the reverse strand. **b** The length of the undigested fragment obtained from S1 nuclease digestion and the length of the reverse transcribed strand were determined by electrophoresing the reaction mixture under denaturing conditions followed by autoradiography. Molecular weight markers were run side by side. *Lane 1* shows the population of reverse transcribed strands and *lane 2* shows the protected fragment from S1 nuclease digestion. The remaining lanes are loaded with sequencing reactions of the GAL1-10 promoter region. Arrow indicates the transcription initiation site (adapted with permission from Johnston & Davis 1984)

such that β-galactosidase transcribed from these constructs would have few codons for N-terminal amino acids of Gal1p or Gal10p. Derivatives of these plasmid were constructed by deletions to varying extent within the *GAL1-10* promoter. The ability of yeast transformant bearing the deletion derivatives and the parent plasmid to express β-galactosidase in response to galactose was monitored (Fig. 5.2.5a). In another independent approach, using the technique of integrative transformation, strains lacking different parts of *GAL1-10* region were constructed. The ability of these strains to express kinase and epimerase enzyme activity was monitored. Results of the above analysis defined an approximately 150-bp GC-rich region located 225 nucleotides upstream from *GAL10* transcription start site and 275 nucleotides upstream from *GAL1* transcription start site. Similar analysis of *GAL7* promoter indicated a GC-rich region in its promoter. Genes coding for permease, phosphoglucomutase and α-galactosidase were later cloned using functional complementation and their promoters were also found to have similar sequences responsible for galactose inducibility.

```
EcoR I
GAATTCGACA GGTTATCAGC AACAACACAG TCATATCCAT TCTCAATTAG CTCTACCACA GTGTGTGAAC CAATGTATCC
CTTAAGCTGT CCAATAGTCG TTGTTGTGTC AGTATAGGTA AGAGTTAATC GAGATGGTGT CACACACTTG GTTACATAGG

            Fok I
AGCACCACCT GTAACCAAAA CAATTTTAGA AGTACTTTCA CTTTGTAACT GAGCTGTCAT TTATATTGAA TTTTCAAAAA
TCGTGGTGGA CATTGGTTTT GTTAAAATCT TCATGAAAGT GAAACATTGA CTCGACAGTA AATATAACTT AAAAGTTTTT

TTCTTACTTT TTTTTTGGAT GGACGCAAAG AAGTTTAATA ATCATATTAC ATGGCATTAC CACCATATAC ATATCCATAT
AAGAATGAAA AAAAAACCTA CCTGCGTTTC TTCAAATTAT TAGTATAATG TACCGTAATG GTGGTATATG TATAGGTATA

CTAATCTTAC TTATATGTTG TGGAAATGTA AAGAGCCCCA TTATCTTAGC CTAAAAAAAC CTTCTCTTTG GAACTTTCAG
GATTAGAATG AATATACAAC ACCTTTACAT TTCTCGGGGT AATAGAATCG GATTTTTTTG GAAGAGAAAC CTTGAAAGTC

TAATACGCTT AACTGCTCAT TGCTATATTG AAGTACGGAT TAGAAGCCGC CGAGCGGGCG ACAGCCCTCC GACGGAAGAC
ATTATGCGAA TTGACGAGTA ACGATATAAC TTCATGCCTA ATCTTCGGCG GCTCGCCCGC TGTCGGGAGG CTGCCTTCTG

TCTCCTCCGT GCGTCCTCGT CTTCACCGGT CGCGTTCCTG AAACGCAGAT GTGCCTCGCG CCGCACTGCT CCGAACAATA
AGAGGAGGCA CGCAGGAGCA GAAGTGGCCA GCGCAAGGAC TTTGCGTCTA CACGGAGCGC GGCGTGACGA GGCTTGTTAT
                                                                  BstN I
AAGATTCTAC AATACTAGCT TTTATGGTTA TGAAGAGGAA AAATTGGCAG TAACCTGGCC CCACAAACCT TCAAATTAAC
TTCTAAGATG TTATGATCGA AAATACCAAT ACTTCTCCTT TTTAACCGTC ATTGGACCGG GGTGTTTGGA AGTTTAATTG

                       Fok I
GAATCAAATT AACAACCATA GGATGATAAT GCGATTAGTT TTTTAGCCTT ATTTCTGGGG TAATTAATCA GCGAAGCGAT
CTTAGTTTAA TTGTTGGTAT CCTACTATTA CGCTAATCAA AAAATCGGAA TAAAGACCCC ATTAATTAGT CGCTTCGCTA

GATTTTTGAT CTATTAACAG ATATATAAAT GGAAAAGCTG CATAACCACT TTAACTAATA CTTTCAACAT TTTCAGTTTG
CTAAAAACTA GATAATTGTC TATATATTTA CCTTTTCGAC GTATTGGTCA AATTGATTAT GAAAGTTGTA AAAGTCAAAC

TATTACTTCT TATTCAAATG TCATAAAAGT ATCAACAAAA AATTGTTAAT ATACCTCTAT ACTTTAACGT CAAGGAGAAA
ATAATGAAGA ATAAGTTTAC AGTATTTTCA TAGTTGTTTT TTAACAATTA TATGGAGATA TGAAATTGCA GTTCCTCTTT
                                  Mbo II
AAACTATAAT GACTAAATCT CATTCAGAAG AAGTGATTGT ACCTGAGTTC AATTCTAGCG CAAAGGAATT ACCAAGACCA
TTTGATATTA CTGATTTAGA GTAAGTCTTC TTCACTAACA TGGACTCAAG TTAAGATCGC GTTTCCTTAA TGGTTCTGGT

TTGGCCGAAA AGTGCCCGAG
AACCGGCTTT TCACGGGCTC
           Ava I
```

Fig. 5.2.4 Nucleotide sequence of the *Eco*RI and *Ava*I fragment. *Arrows* indicate the transcription initiation site. ATG of *GAL1* and *GAL10* are underlined. G residues represented in *boxes* are identified during footprint analysis. Sequences shown in *brackets* are upstream activating sequences to which Gal4p was shown to bind (see section 5.2 for details)

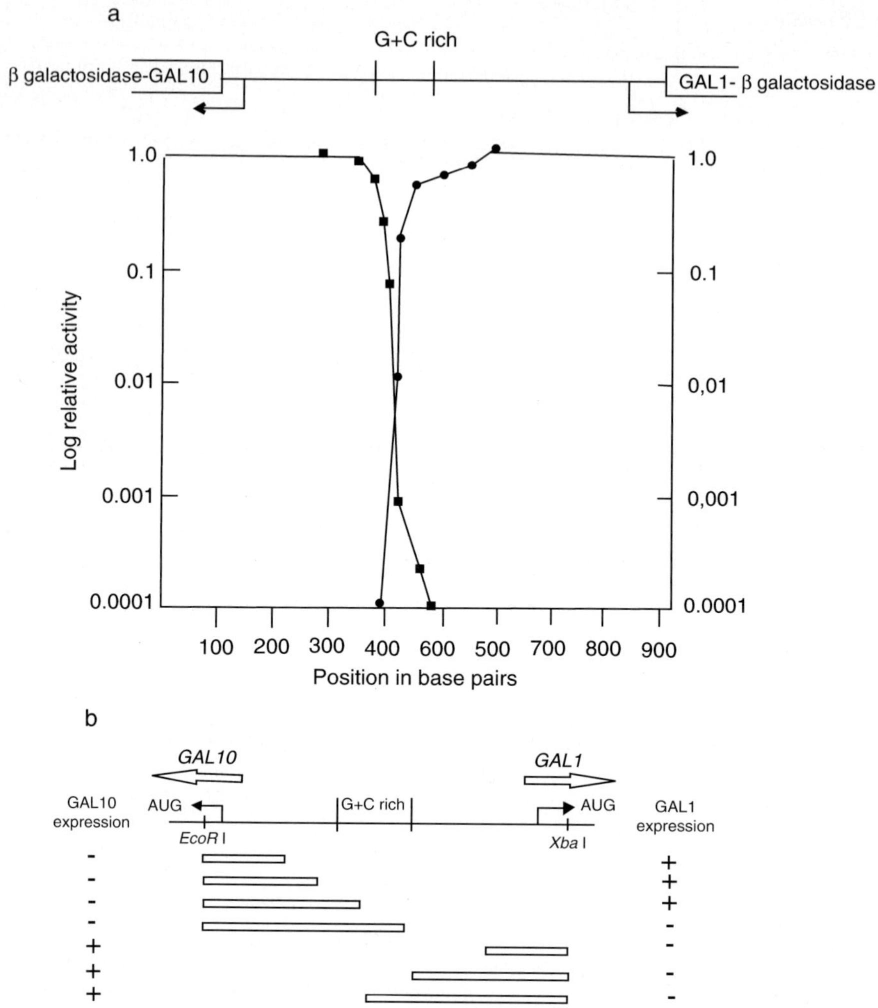

Fig. 5.2.5 Deletion analysis of GAL1-10 intergenic region. **a** β-galactosidase expression driven by GAL1-10 divergent promoter. As the deletion extends into the GC-rich region from *GAL1* or *GAL10* side, β galactosidase activity starts dropping. **b** *Bars* represent the extent of deletion in the genome starting from either the *GAL10* (*EcoR*I end) or *GAL1* (*Ava*I end). + and – indicate the expression status of the respective gene (adapted with permission from Yocum et al. (1984) and Johnston & Davis (1984))

Box 5.2.1 *Galactosemia*

Human galactokinase, uridyl transferase, and epimerase genes have been isolated and studied. Human galactokinase codes for a protein of 420 amino-acid residues. The genes from two individuals clinically diagnosed to have galactokinase deficiency galactosemia revealed the presence of two different mutations. In one case, there is a substitution of V32M and in another case a nonsense codon is introduced at position 238, thus generating a truncated protein. These proteins fail to express galactokinase activity. Availability of human galactokinase sequence allows us to determine whether a given individual carries the defective mutant alleles, which are known to be present in human population. Similarly, mutations in uridyl transferase that cause classic galactosemia have been identified. The most predominant mutation in uridyl transferase is Q188R and K285N with an allele frequency of 69 and 19%, respectively. Either homozygous or compound heterozygous for the above alleles showed late onset complications such as apraxia seizures, tremor, ataxia, dystonia, learning disability, hypogonadism and Mullerian aplasia. Individuals with deficiency in both the copies of epimerase are uncommon.

Box 5.2.2 *Molecular probing*

It is a commonly used technique in biochemistry and molecular biology for the detection and identification of specific proteins and nucleic acids in a heterogeneous population. The sample is first separated in agarose or polyacrylamide gels and transferred to solid support such as nitrocellulose filter paper.

The nitrocellulose paper is then probed with tagged probes to identify the target molecule. If radioactive nucleic acid is used as the probe, then the interaction is driven by sequence similarity, which can be detected by autoradiography. Probes labeled through fluorophores are detected through fluorescence microscopy. Target proteins are identified by using ligands or other interacting proteins. For example, an antibody can be used to detect an antigen present in a mixture of unrelated proteins. In this case, the detection is indirect. Here the antigen antibody complex is detected by a secondary step that involves a reporter molecule fused to an antibody that recognizes the primary antibody. Most often, the reporter molecules are enzymes such as alkaline phosphatase or peroxidase, which permit detection through colorimetric or luminescence assay. In situations where colony or plaques are probed, it is possible to recover the target sample from the master plate as discussed in this chapter. Whole chromosomes can be similarly probed by first fixing the chromosomes onto a glass slide. Whole tissues can be probed for detection of expression of certain protein using immunocytochemical techniques. Here the tissue slices are fixed onto the glass slide then treated with the probe.

References

Guarente L, Yocum RR, Clifford P (1982) A GAL10-CYC1 yeast hybrid promoter identifies the GAL4 regulatory region as an upstream site. Proc Nat Acad Sci USA 79:7410–7414

St. John TP, Davis RW (1979) Isolation of galactose-inducible DNA sequences from *Saccharomyces cerevisiae* by differential plaque filter hybridization. Cell 16:443–452

St. John TP, Davis RW (1981) The organization and transcription of the galactose gene cluster of *Saccharomyces*. J Mol Biol 152:285–315

Johnston M, Davis RW (1984) Sequences that regulate the divergent GAL1-10 promoter in *Saccharomyces cerevisiae*. Mol Cell Biol 4:1440–1448

Oh D, Hopper JE (1990) Transcription of a yeast phosphoglucomutase isozyme gene is galactose inducible and glucose repressible. Mol Cell Biol 10:1415–1422

Post-Beittenmiller MA, Hamilton RW, Hopper JE (1984) Regulation of basal and induced levels of the MEL1 transcript in *Saccharomyces cerevisiae*. Mol Cell Biol 4:1238–1245

Segal S (1998) Galactosemia today: The enigma and the challenge J Inner Metab Dis 21:455–471

Stambolian D et al (1995) Cloning of the galactokinase cDNA and identification of mutations in two families with cataracts. Nature Genetics 10:307–312

Szkutnicka K, Tschoop JF, Andrews L, Cirillo VP (1989) Sequence and structure of the yeast galactose transporter. J Bacteriol 171:4486–4493

West Jr RW, Yocum RR, Ptashne M (1984) *Saccharomyces cerevisiae* GAL1-GAL10 divergent promoter region: location and function of the upstream activating sequence UASg. Mol Cell Biol 4:2467–2478

Yocum RR, Hanley S, West Jr R, Ptashne M (1984) Use of lac Z fusions to delimit regulatory elements of the inducible divergent GAL1-10 promoter in *Saccharomyces cerevisiae*. Mol Cell Biol 4:1985–1988

5.3 Isolation of *GAL4*: The Transcriptional Activator

5.3.1 Introduction

Subtractive hybridization, the approach used to isolate the *GAL* cluster (previous chapter), could not have been used to isolate *GAL4,* as it was predicted to be constitutively synthesized. Instead, it was possible to apply the principle of functional complementation for isolating the genomic fragment corresponding to *GAL4* since a genetically well-characterized haploid yeast strain that has a lesion at the *GAL4* locus was available. Three independent groups reported the isolation of *GAL4* from yeast genomic library.

Fig. 5.3.1 Isolation and analysis of *GAL4*. **a** *gal4 ura3 leu2* mutant strain was transformed to uracil prototrophy using the genomic library. Of the Ura+ transformants, Gal$^+$ transformants were identified by replica plating onto a medium containing galactose as the sole carbon source. Plasmid DNA was isolated from galactose-positive yeast transformants and amplified in *E. coli* for further analysis. **b** Functional region required to complement *gal4* defect was delineated by deletion analysis. The smallest genomic fragment that could complement was found to be in a 3.1*Xba*1 and *Bam*H1 fragment. **c** That this fragment indeed contains *GAL4* was determined using integrative transformation

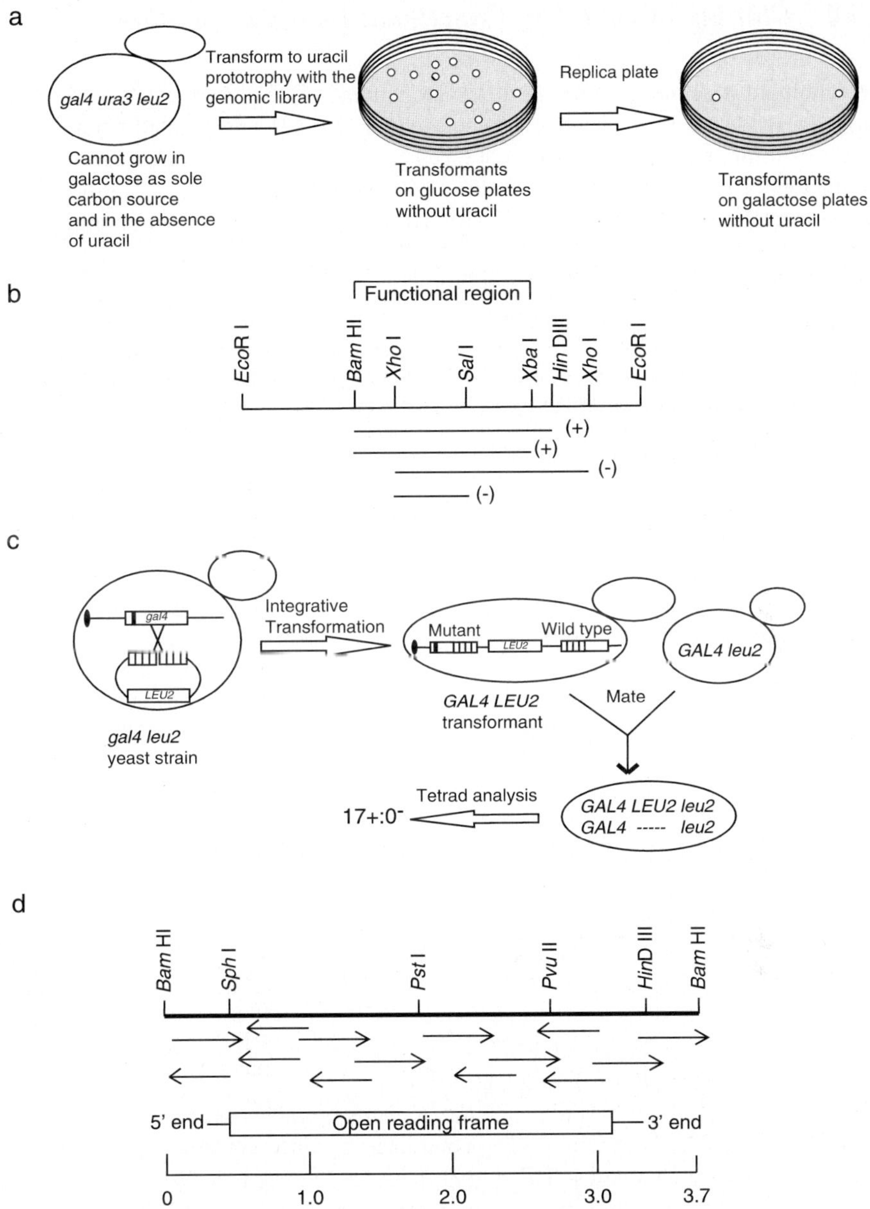

followed by genetic analysis. A 3.1-kb genomic fragment containing *GAL4* was sub-cloned into an integrative plasmid. A *gal4* yeast strain was transformed to galactose prototrophy by integrating plasmid cut with *Xho*1 within the *GAL4* locus to increase the efficiency of integration. The above transformant was mated to *GAL4* strain and the diploids were sporulated and subjected to tetrad analysis. All 17 of the tetrads tested were of the parental combination with respect to growth on galactose, suggesting that the genomic fragment consists of the wild-type *GAL4* allele. **d** Schematic representation of sequencing strategy of 3.7-kb fragment. Both strands of the 3.7-kb length fragment were sequenced. A few *arrows* pointing in both the directions are shown just to indicate that sub-clones of the 3.1 kb were constructed for obtaining the overlapping sequence of both the strands

5.3.2 Cloning of *GAL4* by Functional Complementation

Recombinant plasmid containing a 7.1-kb genomic DNA fragment that complemented a *gal4* strain for growth on galactose (Fig. 5.3.1a) was isolated from a yeast genomic library constructed in YEp24 (see Fig. 5.1.3). The functional region which complements *gal4* defect resides within a 3.1-kb *Bam*H1 *Xho* 1 fragment was shown by deletion analysis (Fig. 5.3.1b). That this genomic fragment corresponds to the genetically defined *GAL4* locus and not any other genomic locus, was demonstrated by carrying out segregation analysis of a diploid formed between a wild-type strain and a *gal4* strain into which the cloned copy of the putative *GAL4* gene was integrated (Fig. 5.3.1c). Of the 17 asci tested, all were parental ditype tetrads with respect to growth on galactose. This indicated that the cloned fragment had integrated at the original *GAL4* locus, suggesting that the cloned fragment corresponds to the cognate *GAL4* gene.

Sequencing of a 3.7-kb fragment containing the functional region of *GAL4* indicated an open reading frame (ORF) of 881 codons, 400 bp of 5′ and 3′ untranslated region. Transcription initiation and termination sites were determined using S1 nuclease mapping. Analysis of the amino-acid sequence revealed that 25% of the first 80 N-terminal amino acids consist of basic amino acids, lysine or arginine and six cysteine residues. It was predicted that these cysteines form disulphide bridges to form a rigid structure with positive charges. A comparison of Gal4p sequence with that of the known bacterial DNA-binding protein sequences revealed structure reminiscent of DNA-binding protein.

5.3.3 *GAL4* Protein Binds Upstream Activating Sequences

Based on (a) genetic analysis indicating that *GAL4* is essential for activation of *GAL* genes (Chap. 3.1) (b) the presence of DNA sequence elements on the *GAL* promoter confer galactose inducibility (Chap. 5.2) (c) similarity of Gal4p to other known DNA-binding protein, it was speculated that Gal4p is a DNA-binding transcriptional activator. Availability of *GAL4* on a multicopy plasmid made it possible to assay for DNA-binding activity.

Evidence that Gal4p binds to promoter elements of galactose inducible genes has come from many independent studies conducted by different groups. Here I shall discuss only two such studies. The region that is recognized by Gal4p in *GAL1-10* promoter was demonstrated by in vivo footprinting using methylation protection assay. This approach identified four regions protected by Gal4p (Fig. 5.3.2). The sequences identified using the in vivo footprinting assay is in agreement with the sequence elements identified using the deletion analysis of GAL1-10 promoter (Fig. 5.2.5). Thus, it was inferred that in a wild-type strain growing in a non-inducing medium, Gal4p remains bound to the specific sequences referred to as upstream activating sequences (UASg), but unable to activate transcription probably due to

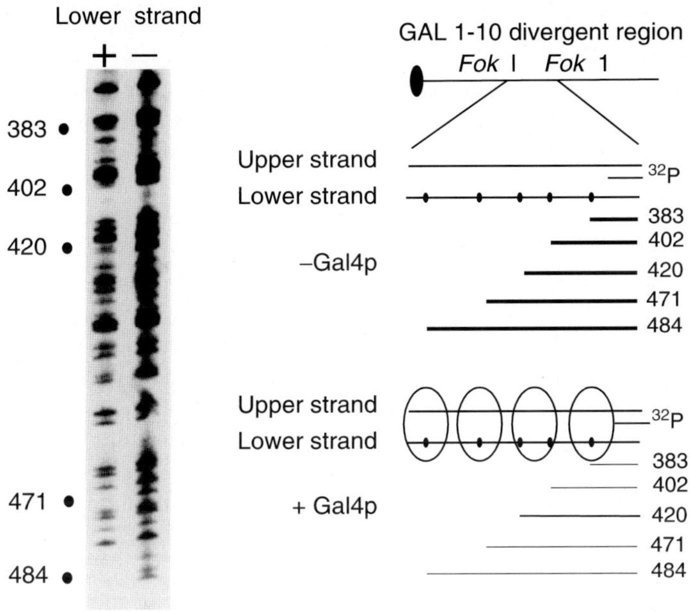

Fig. 5.3.2 In vivo footprinting by methylation using dimethyl sulphate. Yeast cells with (+) or without (−) *GAL4* were grown in non-repressing carbon source and the cells were treated with dimethyl sulphate (DMS). DMS adds methyl group primarily to A and G bases located in the minor and major grove, respectively, and makes the DNA susceptible to cleavage by piperidine. The presence of bound proteins changes the methylation pattern. The difference in the methylation pattern was detected after isolating the chromosomal DNA followed by digestion with *Fok* 1 restriction enzyme (two *Fok*I sites are present in GAL1-10 region see, Fig. 5.2.4) and cleave the DNA with piperidine. Fragments are then separated through denaturing PAGE, transferred to nitrocellulose and hybridized with strand-specific single-strand 32p labeled DNA probe corresponding to GAL1-10 region. Position 383, 402, 420, 471, and 484 represent the position of the G residue protected in DNA isolated from cells expressing GAL4 (intensity is less), while the corresponding positions are not protected from cells not expressing GAL4 (intensity is more) and therefore cleaved by piperidine. The *right panel* shows the schematic representation of the principle of methylation assay. Open circles represent Gal4p while G residues are indicated by small filled circles (adapted with permission from Giniger et al. 1985)

the inhibition conferred by Gal80p, the repressor. In another independent study, protein fraction enriched for Gal4p, obtained from yeast strains over-expressing Gal4p (Box 5.3.1), protected two regions in *GAL7* promoter from DNase I digestion (Fig. 5.3.3). The length of the protected region was of the order of 30 bp separated by 55 bp. These preliminary results indicated that Gal4p recognizes specific region in *GAL7* promoter as well.

Later experiments demonstrated that Gal4p protein binding sites are present in promoters of known galactose inducible genes. Extensive biochemical and mutational analysis indicated that the Gal4p binds to a 17-bp consensus sequence 5′ CGG N_5

Box1 5.3.1 *Filter binding assay*

This technique was originally developed by Arthur Riggs and his collaborators to detect the binding of Lac repressor with its operator. This was based on the observation that proteins (but not double-stranded DNA) bind to nitrocellulose filter paper. This property is exploited in separating the protein DNA complex from a reaction mixture containing the free protein, DNA and the complex. The DNA fragment is radioactively labeled and allowed to interact with varying concentration of DNA-binding protein. The extent of binding is monitored by counting the radioactivity retained on the nitrocellulose filter disc. The radioactive counts retained on the filter paper is a function of the concentration of DNA-binding protein. Before this technique, the DNA protein complexes were separated from free DNA and protein by centrifugation through glycerol density gradients, which was laborious and time-consuming. Filter-binding assay was used to monitor the activity of Gal4p protein in cell extracts obtained from yeast strains bearing multicopy GAL4 expression plasmids.

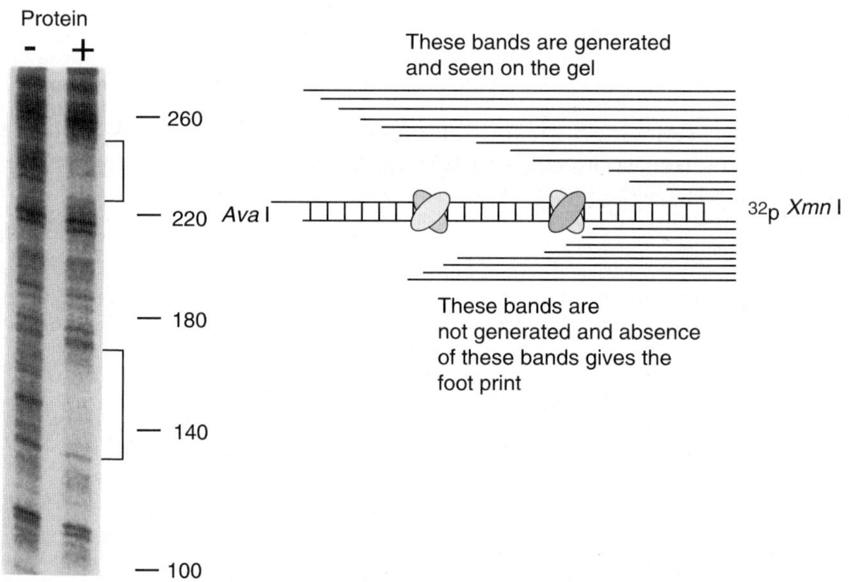

Fig. 5.3.3 Dnase I footprinting of *GAL7* promoter. A DNA fragment extending from *Xmn*1 site at –405 (+1 represents GAL7 ATG) to *Ava*I site at a position within GAL7 coding region was labeled at *Xmn*I end with 32p. This fragment was partially digested with Dnase I in the presence or in the absence of the partially enriched extract containing Gal4p. Dnase I cleaves at random between any two successive nucleotide residues except where the DNA is protected by the protein. After the reaction, DNA is subjected to electrophoresis under denaturing conditions, which results in a series of radioactive single-stranded fragments extending from the labeled end to each position in the chain except within the Gal4p-binding region. The sizes of the missing fragments corresponding to the distances from the labeled 5′end of the fragment are determined by running a sequencing ladder (adapted with permission from Bram and Kornberg 1985)

T N_5 CCG 3′ which has a two-fold rotational symmetry. The CGG triplets are on the opposite sides of the helix separated by one and a half turn. Detailed mutational analysis has indicated that mutations in the second base pair of one or both of the CGG triplet severely inhibits binding of Gal4p. Changing the central base pair AT to GC does not have any affect on binding. These sequences function in either direction and at various distances from the gene. Later, it was shown that just one UASg element is sufficient to confer galactose inducibility. In fact, *MEL1* was shown to have only one UASg in its promoter.

5.3.4 GAL4 Protein Binds GAL80 Protein

Availability of cloned GAL4 gene, provided a convenient handle to test the following predictions of the of protein–protein interaction model. That is (a) *GAL4* transcription is constitutive (b) Gal80p directly interacts with Gal4p to inhibit transcriptional activation function of Gal4p. The first prediction was tested in hybridization experiments where in ^{32}P labeled DNA fragment containing *GAL4* was used as a probe. This probe hybridized to a 2.8-kb mRNA isolated from cells grown in glycerol, galactose but not glucose, indicating that GAL4 mRNA is constitutively transcribed in glycerol and galactose and not expressed in glucose-grown cells. This observation supported the hypothesis that galactose does not induce the expression of Gal4p, which is contrary to what Douglas and Hawthorne had proposed. This also indicated that glucose represses *GAL* gene expression by inhibiting transcription of *GAL4*. Northern blot analysis also indicated that in comparison to other known genes, *GAL4* mRNA was not abundant, suggesting that it is a weakly expressed gene. This is anticipated as it is expected that regulatory proteins are normally not highly expressed.

In an independent study, over-expression of Gal4p from a multicopy plasmid, conferred constitutive expression of *GAL* genes not only in wild-type strain but in strain which is otherwise non-inducible due to the presence dominant *GAL80*$^{s-2}$ allele (recall that *GAL80*$^{s-2}$ is epistatic over wild-type *GAL4* allele). Over-expression of a truncated Gal4p lacking 50 N-terminal amino-acid residues also conferred constitutive expression of *GAL* genes in a wild-type and *GAL80*$^{s-2}$ strain. Surprisingly, unlike the wild-type, the N-terminal truncated Gal4p did not complement a *gal4* strain for growth on galactose. These results pointed out (a) wild-type and N-terminal truncated Gal4p probably titrate GAL80 and GAL80^{s-2} protein by direct interaction resulting in wild-type Gal4p being free to constitutively activate *GAL* genes. (b) N-terminal truncation of Gal4p leads to the inability to activate *GAL* genes in *gal4* strain because of lack of the DNA-binding domain. But the fact that the N-terminal truncated Gal4p suppresses the epistatic effect of *GAL80*$^{s-2}$ suggested the C-terminal of Gal4p has Gal80p-binding domain. Based on these results, it was inferred that the N-terminal part codes for DNA-binding domain while the C-terminal part codes for Gal80p-binding domain.

Direct evidence that Gal4p interacts with Gal80p came from many independent studies. Gal4p was purified by employing conventional protein fractionation steps from yeast cell-free extracts obtained from Gal4p over-expressing strain (Fig. 5.3.4). Two independent laboratories consistently observed that a protein of molecular

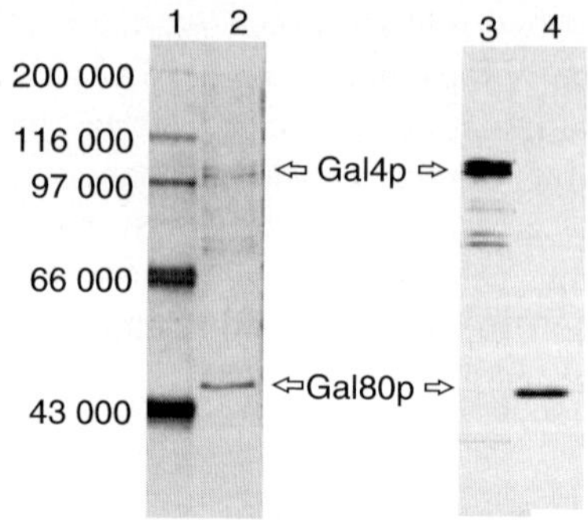

Fig. 5.3.4 Over-expression and purification of Gal4p. Gal4p expression was induced by adding galactose to wild-type yeast transformants bearing a multicopy plasmid containing GAL10:GAL4 expression cassette. Cell-free protein extract obtained from yeast after induction was fractionated using different chromatographic steps. SDS electrophoresis pattern of active fraction obtained after affinity purification through UASg DNA sepharose columns, stained using Coomassie blue (*lane 2*). *Lane 1* represents molecular weight markers. Western blot analysis of the protein fraction using antibodies raised against GAL4 (*lane 3*) and GAL80 protein (*lane 4*) (adapted with permission from Chasman and Kornberg 1990)

weight 48 kDa gets co-purified with Gal4 protein (Fig. 5.3.4). The 48-kDa protein was later identified as Gal80p by Western blot analysis. These findings provide direct evidence that Gal80p is a repressor that directly interacts with Gal4p to inhibit its transcriptional activation function in the absence of galactose. That Gal80p is co-purified with Gal4p even when cell-free extracts prepared form galactose grown cells suggested that the induction signal may not dissociate the Gal4p-Gal80p complex.

5.3.5 *GAL4 Protein is Modular*

Gal4p has been subjected to exhaustive mutational and structural analysis. Because of the large size of Gal4p, most analyses have been carried out on specific domains of Gal4p either as independent entities or as fusion proteins. These studies revealed that GAL4 is a modular protein with separable DNA-binding and transcriptional activation domains.

DNA-binding domain: DNA-binding activity resides in the first 74 amino acids of Gal4p was demonstrated by the following experiment. First, 74 amino acids of Gal4p were replaced by 87 amino-acid residues of bacterial DNA-binding repressor, lex A. It was previously known that 87 amino-acid residues of lex A are sufficient to recognize lexA operator in *E. coli*. This hybrid protein did not complement a *GAL4* defect of a yeast strain for growth on galactose. However, over-expression of this hybrid protein in a wild-type strain titrated the Gal80p repressor, allowing the wild-type Gal4p to activate the *GAL1* promoter. This hybrid protein did activate the transcription from GAL1 promoter only when the UASg elements were replaced by the *E. coli* lexA operator. These results corroborated that the amino-acid residues 87 to 881 of Gal4p not only activate transcription but also recognize GAL80 protein, an observation reminiscent of the observation discussed in previous section. Another important implication from the above experiment was that the function of DNA-binding domain is to recruit the transcription activating region of GAL4 protein to the appropriate site and DNA binding per se has no role in transcription activation. Most likely the transcription activation is carried out by the interaction of the C-terminal part of Gal4p with the transcriptional machinery. These initial observations paved the way for a detailed analysis of the DNA-binding region of Gal4p.

Exhaustive mutational analysis provided genetic evidence that GAL4 is a DNA-binding transcriptional activator protein with the DNA-binding domain consisting of a zinc finger. A total of 88 independent mutations that abolished Gal4p function were isolated; 37 out of a total of 46 mutations that affect DNA binding, occurred between amino-acid residues 10 to 57 (Fig. 5.3.5). Out of the 42 mutations that lie in regions

a

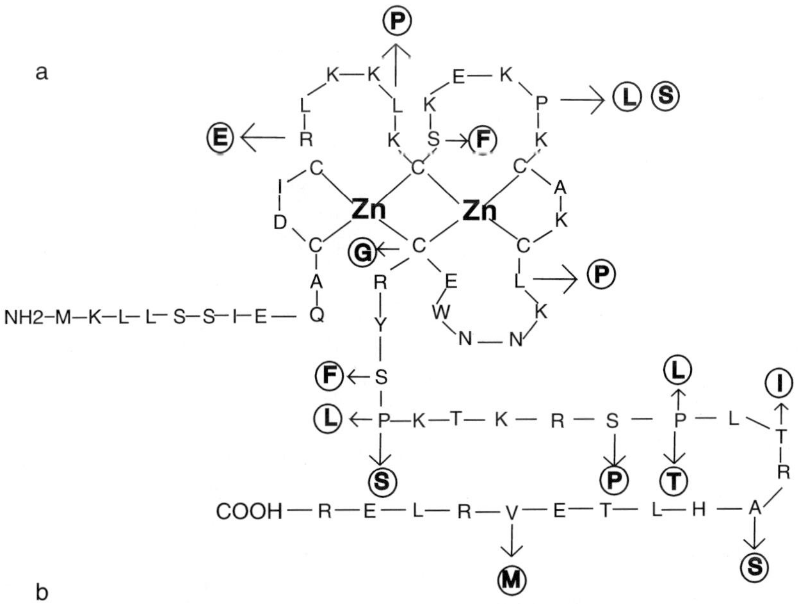

b

GAL4 ....A Y N F G I T T G M F N T T T M D D V Y L F D D E D T P P N P K K E

gal4.62...... A Y N F G I T T End

GAL4^c.62.......A Y N A F M N V End

Fig. 5.3.5 N-terminal DNA binding and C-terminal transcription domain. **a** DNA-binding domain forms a binuclear metal cluster involving four cysteines in a cloverleaf-like fashion. Substitution of some important residues that leads to loss of DNA-binding domain are indicated. **b** The amino-acid sequence of the C-terminal end of wild-type, *gal4.62* and *gal4^C.62* proteins is shown

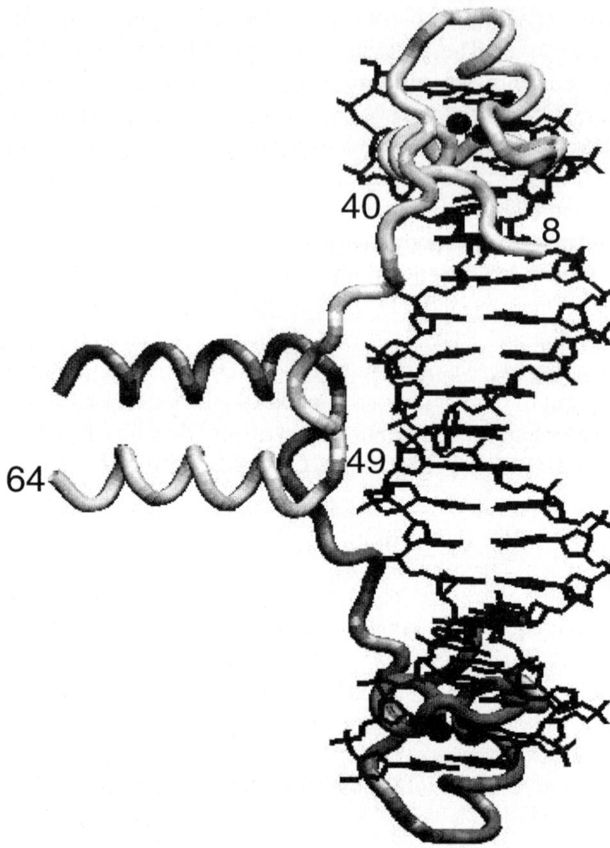

Fig. 5.3.6 DNA-binding domain of Gal4p. Schematic representation of interaction between N-terminal 1–65 amino acids of GAL4 protein and 17-bp UASg element. The complex is viewed approximately perpendicular along its two-fold axis of symmetry. Amino-acid residues 8–40 is the DNA recognition module, 41–49 is the linker region and 50–64 is the dimerization region (picture drawn based on the coordinates obtained from Marmorstein et al. 1992)

spanning from amino-acid residues 75 to 881, only three are missense mutations that affect transcription and not DNA binding. The rest of the mutations were of either deletion or nonsense type (generally misense mutant derivatives are more informative in terms of structure activity relationship, while deletion mutants provide insights into domain structure). That the inability of a P26L Gal4p mutant to complement Gal^- phenotype of *gal4* cells could be alleviated by increasing the Zn ion concentration in the growth medium provides evidence for the biological role for the zinc fingers in DNA binding. These results suggest that Gal4p contains a functional metal-binding DNA-binding domain. The DNA-binding domain consisting of 1-147 residues binds the consensus UASg elements in vitro in presence of Zn, Cd or Hg.

Box 5.3.2 *Reverse genetics*

This term was coined by Charles Weismann before the advent of recombinant DNA technology. It involves the analysis of new phenotypes induced by directed changes in the structure of DNA. Classical genetics involves analysis of naturally occurring or induced mutant organisms. The principle aim of classical genetics was to identify genetic loci, map them, and mutate them to observe their phenotypic affect. For example, the inability to express Leloir genes was shown to be due to a defect at the *GAL4* locus. Using mapping techniques it was shown to be located on chromosome XVI. On the other hand reverse genetics begins with the availability of the wild-type cloned gene. For example *GAL4* gene is altered in vitro and its effects are monitored by transferring the modified gene back into the organisms to study its mechanism of action. The starting point for reverse genetics is a gene as a piece of DNA.

Box 5.3.3 *DNA-biding motifs*

Transcription activators recognize short nucleotide sequence because of the structural complementarity between the surfaces features of protein and the specific DNA sequences. Although the protein DNA complex is stabilized by weak interaction such as hydrogen bonds, ionic interactions or hydrophobic interactions, they collectively ensure that the binding is specific and strong. The DNA-binding domain of many activators has been well studied and a few conserved structural motifs have been identified that are common to different transcription factors with quite different specificities. In addition to the zinc finger domain, the leucine zipper, helix loop helix, and helix turn helix are the most commonly found DNA-binding domains. Each of the motifs uses α- helix to bind to the major groove of DNA. In leucine zipper, monomeric units consist of an amphipathic α-helix with leucine residues spaced seven amino acids so that they are all present on one face of the helix. Two such monomers join in a zipper-like fashion to form a coil. The overall dimer is a Y-shaped structure and the arms of the Y recognize the major groove. Leucine zipper can also occasionally form heterodimers, thus providing a combinatorial control in gene expression. An example of the leucine zipper protein is the GCN4 protein, a transcriptional activator of amino acid biosynthetic genes in *Saccharomyces cerevisiae*. Helix loop helix domain consists of one short and one long helix connected by a flexible loop. Just like the leucine zipper motif, the HLH motif mediates both DNA binding and dimerization. In contrast, the helix turn helix proteins contain two helices, two of which are short and are separated by short amino-acid sequences, which induces a turn and constitutes the helix turn helix motif. This structure is very similar to the DNA-binding motif of many bacterial repressors and present in homeobox domains of eucaryotic transcriptional activators.

Gal4p (1-65) form exists as a monomer in solution in the absence of DNA.Gal4p (1-149) dimerizes in solution and bind DNA in a cooperative manner with >1,000-fold higher affinity than the Gal4(1-65). It has also been shown that binding of this fragment bends DNA, which is implicated to have a role in the transcriptional activation. Based on structural analysis of a consensus UASg/Gal4p complex consisting of amino-acid residues from 1-65, the binding domain is localized from residues 14 to 57 and this module is shown to be a C6 zinc cluster. Here, two zinc ions are coordinated to six cysteine residues with the remaining amino acids folded around the Zn_2Cys_6 forming a bimetal thiolate cluster. Gal4p binds UASg as dimer with each monomer recognizing one-half of the dyad symmetric-binding site. A dimer interface is a parallel coiled coil formed by amphipathic helices of residues 50-64. A linker formed by amino-acid residues 41–49 connects the zinc cluster to the dimer interface. Amino-acid residues of the linker and the coiled coil contact the DNA backbone within the UASg. Each C_6 zinc cluster makes base-specific contacts with the terminal CCG triplet, which are the most conserved triplets in UASg. The region necessary for the nuclear localization is present in the first 74 amino acids and overlaps with the DNA-binding region, but these two functions are mutationally separable.

Gal4p is not the only Zn_2Cys_6 class of proteins present in yeast which recognize CGG triplets. For example, PPR1, and PUT3 proteins, involved in the regulation of pyrimidine and proline metabolism, also belong to the Zn_2Cys_6 class and recognize CGG triplets but do not activate *GAL* genes. If so, what features of Gal4p and the UASg impart the specificity? One discriminatory feature is the number of central base pairs present in the UASg elements. For example, Ppr1p recognizes sites with 6-bp spacing while Gal4p binds with 11-bp spacing. Domain-swap experiments wherein C_6Zn_2 finger of Gal4p replaced by corresponding part of Ppr1p, binds UASg elements and activates transcription of *MEL1* gene. On the other hand, substituting the dimerization domain of Gal4p with Ppr1p dimerization domain binds the Ppr1p-binding site. These experiments indicate that the dimerization domain plays a critical role in providing specificity. In addition, under in vivo conditions, association with other proteins could also impart higher specificity.

Transcriptional activation domain: The transcriptional activation domain was also extensively analyzed both by structural and mutational analysis. Recall, that Oshimas group had isolated a non-functional nonsense allele *gal4.62* from which a constitutive allele *GAL4*^{C}.62 was isolated. The mapping of the mutant site indicated that it is close to the site of the nonsense mutation (see Chap. 3.1). Cloned *GAL4* gene was used to retrieve the *gal4.62* and *gal4*^{C}.62 alleles from the corresponding mutant haploid strains. Sequencing of these mutant alleles revealed that in *gal4.62*, codon 854 was converted to TGA, a nonsense codon (Fig. 5.3.5). The constitutive revertant *gal4*^{C} *-62* allele, has an 11-bp deletion just upstream of codon 854. Frame shift caused by 11-bp deletion resulted in a protein of 852 amino acids, of which 849 amino acids are normal and the last three are Met-Asn-Val. While *gal4.62* is transcriptionally inactive, gal4^C -62 protein restored the activation function without acquiring the ability to interact with Gal80p. These results are in agreement with the previous results that C-terminal sequence is responsible for Gal80p recognition.

In another independent study, deletion of the C-terminal 30 amino acids conferred constitutive induction, indicating that these residues interact with Gal80p.

In addition to the above, amino-acid segments 148–196 fused to the DNA-binding region activate transcription (activating region I). Both are prototypical acidic activation domains but it is the C-terminal transcription activation region (activating region II) that is inhibited by the Gal80p. Presence of acidic activation domain is a common feature of many eucaryotic transcriptional activators. Although the region II has both transcription activating and GAL80-binding activity, these two functions are mutationally separable. For example, the $GAL4^c$ alleles activate transcription but are not antagonized by GAL80 protein. In several of these alleles, single amino-acid changes were found in residues spanning 859–868.

The following experiment illustrates the relationship between the DNA-binding and transcriptional activation region. A *gal4gal80* strain transformed with a plasmid construct consisting of a 1–147 and 851–881 weakly activates transcription, but this chimeric protein interacts with the Gal80 protein due to the presence of the Gal80p-binding region consisting of amino-acid segment 851–881. The above transformant activated the *GAL1* promoter when transformed with a second plasmid construct consisting of *GAL80* into which an acidic transcription activating region is inserted. This experiment clearly indicated that the DNA binding and transcription functions need not necessarily reside in the same protein for gene-specific activation and that the transcription activating sequence brought to the appropriate site through non-covalent interaction is sufficient to activate transcription.

Detailed mutagenesis of the polypeptide region extending from 855 to 870 indicated that it is relatively insensitive to mutations with regard to transcriptional activation but not Gal80p binding function. It was shown that the region extending from 840 to 874 is unstructured in solution at physiological pH but formed a β-hairpin structure at pH 5.9. Polypeptide region extending from amino-acid residues 855 to 870 is sensitive to mutational perturbation with regard to Gal80p binding but not in activating transcription or binding to TATA-binding protein. Moreover, Gal80p-binding peptides isolated form peptide library, activate transcription when fused to the DNA-binding domain. Thus the structure of transcription activating domain appears to be more promiscuous than the Gal80p-binding structure.

References

Ansari AZ, Reece RJ, Ptashne M (1998) A transcriptional activating region with two contrasting modes of protein interaction. Proc Nat Acad Sci USA 95:13543–13548

Berg P (1991) Reverse genetics: Its origin and prospects. Biotechnology 9:342–344

Bram RJ, Kornberg RD (1985) Specific protein binding to far upstream activating sequences in polymerase II promoters. Proc Nat Acad Sci USA 82:43–47

Brent R, Ptashne M (1985) A eukaryotic transcriptional activator bearing the DNA specificity of a prokaryotic repressor. Cell 43:729–735

Chasman D, Kornberg, RD (1990) GAL4 protein: purification and association with GAL80 protein and conserved domain structure. Mol Cell Biol 10:2916–2923

Giniger E, Varnum SM, Ptashne M (1985) Specific DNA binding of the GAL4: a positive regulatory protein of yeast. Cell 40:764–774

Han Ying, Kodadek T (2000) Peptides selected to bind the GAL80 repressor are potent transcriptional activation domains in yeast. J Biol Chem 275:14979–14984

Hishimoto H, Kikuchi Y, Nogi Y, Fukasawa T (1983) Regulation of expression of the galactose gene cluster in *Saccharomyces cerevisiae*. Mol Gen 191:31–38

Johnston M (1987) A model fungal gene regulatory mechanism: the *GAL* genes of *Saccharomyces cerevisiae*. Microbiol Rev 51:458–476

Johnston M (1987) Genetic evidence that zinc is an essential co-factor in the DNA-binding domain of Gal4p. Nature 328:353–355

Johnston SA, Hopper JE (1982) Isolation of the yeast regulatory gene GAL4 and analysis of its dosage effects on the galactose/melibiose regulon. Proc Nat Acad Sci USA 79:6971–6975

Johnston M, Dover J (1987) Mutations that inactivate a yeast transcriptional regulatory protein cluster in an evolutionary conserved domain. Proc Nat Acad Sci USA 84:2401–2405

Johnston M, Dover J (1988) Mutational analysis of the GAL4 encoded transcriptional activator protein of *Saccharomyces cerevisiae*. Genetics 120:63–74

Johnston SA, Salmeron JM, Dincher Jr SS (1987) Interaction of positive and negative regulatory proteins in the galactose regulon of yeast. Cell 50:143–146

Johnston SA, Zavortink MJ, Hopper JE, Debouck C, Hopper JE (1986) Functional domains of the yeast regulatory protein GAL4. Proc Nat Acad Sci USA 83:6553–6557

Keegan L, Gill G, Ptashne M (1986) Separation of DNA binding from the transcription activating function of a eukaryotic regulatory protein. Science 231:699–703

Laughon A, Gesteland RF (1982) Isolation and preliminary characterization of the GAL4 gene, a positive regulator of transcription in yeast. Proc Nat Acad Sci USA 79:6827–6831

Laughon A, Gesteland RF (1984) Primary structure of the *Saccharomyces cerevisiae* GAL4 gene. Mol Cell Biol 4:260–267

Lohr D, Hopper JE (1985) The relationship of regulatory proteins and DNAse I hypersensitive sites in yeast GAL1-10 genes. Nucleic Acids Res 13(23):8409–8423

Lohr D, Venkov P, Zlatanova J (1995) Transcriptional regulation in the yeast *GAL* gene family: a complex genetic network. FASEB J 9:777–786

Ma J, Ptashne M (1987) The carboxy terminal 30 amino acids of GAL4 are recognized by GAL80. Cell 50:137–142

Ma J, Ptashne M (1988) Converting a eukaryotic transcriptional inhibitor into an activator. Cell 55:443–446

Marmorstein R, Carey M, Ptashne M, Harrison C (1992) DNA recognition by GAL4: structure of a protein-DNA complex. Nature 356:408–414

Melcher K (1996) Galactose metabolism in *Saccharomyces cerevisiae*: a paradigm for eukaryotic gene regulation. In: Zimmermann K, Entian KD (eds) Yeast sugar metabolism. Technomic Publishing Co., Lancaster, PA

Parthun M, Jaehning JA (1992) A transcriptionally active form of GAL4 is phosphorylated and associated with GAL80. J Biol Chem 12: 4981–4987

Ptashne M (1992) A genetic switch: phage lambda and higher organisms, 2nd edn. Cell Press & Blackwell Scientific Publications, Cambridge

Vashee S, Xu H, Johnston SA, Kodadek T (1993) How do "Zn2 Cys" proteins distinguish between similar upstream activation sites. J Biol Chem 268:24699–24706.

Riggs AD, Suzuki H, Bourgeosis S (1970) Lac repressor operator interactions I. Equilibrium studies. J Mol Biol 48:67–83

Selleck SB, Majors JE (1987) In vivo DNA-binding properties of a yeast transcription activator. Mol Cell Biol 7:3260–3267

Salmeron JM Jr, Leuther KK, Johnston SA (1990) GAL4 mutations that separate the transcriptional activation and GAL80 interactive functions of the yeast GAL4 protein. Genetics 125(1):21–27

5.4 Isolation of *GAL80*: The Repressor

5.4.1 Introduction

Gal4p remains bound to UASg when cells grow in a non-inducing and non-repressing carbon source, but Gal4p bound to DNA is unable to activate transcription. Experiments discussed thus far point out that Gal80p prevents Gal4p from activating transcription by direct interaction. Gal3p in the presence of galactose through mechanisms not clearly understood, negates the inhibition conferred by Gal80p, thus allowing Gal4p to activate transcription. Molecular details of many of these steps are not understood. Isolation of *GAL80* proved crucial in elucidating the mechanistic basis of Gal80p inhibition of Gal4p.

5.4.2 Cloning of GAL80 by Genetic Suppression

GAL80 was isolated from a multicopy genomic library based on the phenomenon of genetic suppression. In the presence of galactose, accumulation of galactose 1-phosphate causes toxicity due to the absence of uridyl transferase, and therefore prevents the cell from using alternate carbon sources such as ethanol. Being a repressor, over-expression of *GAL80* was expected to suppress this phenotype because of the transcriptional repression of galactokinase. Based on this principle (Fig. 5.4.1), a recombinant plasmid bearing a 4.5-kb genomic fragment containing the wild-type *GAL80* allele was isolated from the genomic library. That this fragment corresponds to the genomic locus in which *GAL80* locus resides was confirmed by using a strategy similar to the one discussed in the previous section. Deletion analysis narrowed the functional region to be within 3.0-kb fragment present within *Hin*DIII and *Xho*I sites. Using the strategy discussed above, *GAL80*^{s0}, *GAL80*^{s1}, *GAL80*^{s2} alleles were also isolated from genomic DNA library prepared from total genomic DNA isolated from the respective strain.

Sequence of 2,457 bp of this fragment contained an open reading frame of 1,275 bp, which codes for a protein of 435 amino acids. S1 mapping indicated that the transcription initiation occurs 75 bp 5′ to ATG and termination occurs at 100 bp 3′ to the termination codon. Accordingly, a 1.55-kb-long mRNA was detected with Northern blot analysis using the *GAL80* gene as a probe. Amino-acid sequence of the Gal80 protein did not reveal any unusual features. Lack of an in vitro assayable biochemical activity precluded purification of Gal80p from cell-free extract. Antibodies raised against a synthetic peptide corresponding to amino-acid residues 247 to 264, reacted specifically with a protein of expected molecular weight only from cell extract obtained from a Gal80p over producer. Using immunoblot or Western blot analysis as an assay, Gal80p was purified from cell extract using a variety of fractionation steps. That purified Gal80p interacts with Gal4p was demonstrated using a gel retardation assay (Box 5.4.1). It revealed that Gal80p can also

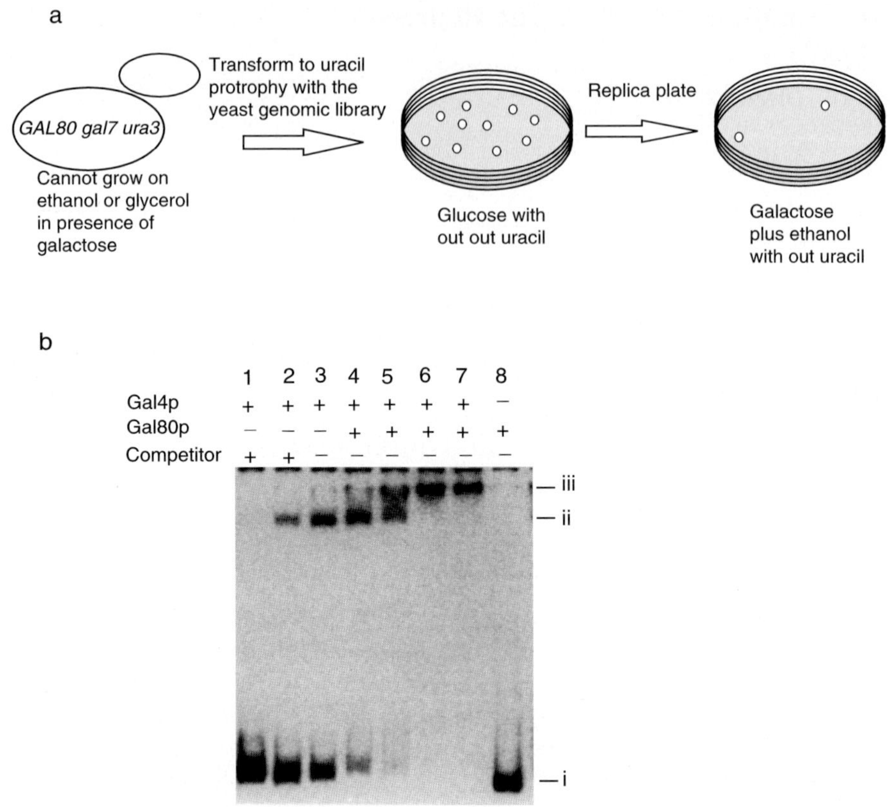

Fig. 5.4.1 *GAL80* cloning strategy. **a** A *GAl80 gal7 or gal80 gal7* mutant cannot utilize alternate carbon source such as ethanol in the presence of galactose due to the accumulation of galactose 1-phosphate, which is toxic. Suppression of galactokinase synthesis is expected to relieve this toxicity, which serves as the convenient selection protocol for isolating transformants that bear multiple copies of *GAL80*. **b** Gel-shift assay demonstrating the Gal4p-binding activity of the *GAL80* protein. ^{32}P labeled 416-bp *Fok*I fragment of GAL1-10 promoter is the source of UASg. A total of 5 μg of partially purified Gal4p and purified Gal80p ranging from 5,10,15,25, 25, ng were used. Position of free DNA (i) DNA-Gal4p (ii) and DNA-Gal4p-Gal80p complex (iii) is indicated (adapted with permission from Yun et al. 1991).

interact with Gal4p bound to DNA (Fig. 5.4.1b). Gal80p alone did not interact with the labeled DNA, clearly indicating that the repression is mediated through a direct protein–protein interaction between Gal4p and Gal80p.

5.4.3 Autogenous Regulation of *GAL80* Expression

Northern blot analysis revealed that its transcription is induced five-fold upon galactose addition. Consistent with this, the *GAL80* promoter contains a functional 17-bp consensus Gal4p-binding UASg element. Detailed analysis revealed that

Box 5.4.1 *Gel-shift analysis*

In this technique, ^{32}P labeled target DNA fragment is allowed to form a complex with the DNA-binding protein. The protein DNA complex is separated from the free-labeled DNA by allowing these species to migrate in a polyacrylamide gel under the influence of an electric field. Since the protein DNA complex is larger than the DNA, the complex moves slower. The position of both the free DNA and the protein DNA complex is detected by autoradiography. This technique can also be used to demonstrate the interaction of DNA-binding proteins with other proteins as in the case of Gal4p. In this case, the DNA-Gal4p-Gal80p complex is larger than just the DNA-Gal4p complex and is retarded further. This powerful technique was independently developed by two groups: Garner and Revzin as well as Fried and Crothers.

Box 5.4.2 *Autogenous regulation*

The original concept of gene regulation by regulatory genes proposed by Jacob and Monod obviously raised the following question. That is, what regulates the expression of regulatory proteins? A protein specified by a gene regulates its own expression, in addition to the regulation of expression of other genes that are normally under its control. That is, the regulatory protein regulates the rate at which additional copies of the same protein are synthesized. One of the classic examples is that of the positive effect of lambda repressor on its own gene cI. Autogenous regulation is a sophisticated mechanism that has many ramifications in a variety of biological manifestations.

transcription of *GAL80* is regulated by the same switch that it is a part of. This phenomenon is referred to as autogenous regulation (Box 5.4.2). It was initially thought that autoregulation is a mechanism to ensure that the cells do not synthesize enzymes more than what is needed by the cell. However, it is now becoming clear that autoregulation of regulatory proteins through feedback mechanisms imparts unique properties to the genetic circuit. (see section 8.2.4).

5.4.4 Mutational Analysis of *GAL80*

Gal80p is a relatively small but is expected to have many functional domains. The sequencing of super-repressor alleles $GAL80^{s\,0}$, $GAL80^{s\,1}$ and $GAL80^{s\,2}$ indicated that they bear G310R, G323R and E351K substitutions localized to a narrow stretch of

50 amino acids. Deletion analysis also indicated that this region is responsible for inducer interaction although at that time the nature of the inducing signal was not known. Observation that Gal80p can bind Gal4p bound to the DNA raised the possibility that it enters the nucleus. Sub-cellular location of Gal80p was indirectly monitored by observing the distribution of β-galactosidase to which various parts of Gal80p are fused. β-galactosidase distribution within the cell was monitored using an indirect immunofluorescence staining technique. This analysis indicated that amino-acid residues from 1–109 or from 342–405 are capable of directing the β-galactosidase synthesized in the cytoplasm to the nucleus.

Most cells express bleomycin hydrolase, an enzyme that renders bleomycin unable to cleave DNA. Bleomycins are a family of glycopeptides produced by *Streptomyces verticillus*. It was fortuitously observed that bleomycin hydrolase was induced several-fold when yeast cells grow on galactose as the sole carbon source but is repressed on glucose, a feature reminiscent of the *GAL* genes. Accordingly, a functional UASg was identified in its promoter and, as expected, galactose-dependent induction of bleomycin hydrolase is dependent on Gal4p. Based on these features, bleomycin hydrolase is considered as yet another member of *GAL* gene family and is called *GAL6*. A surprising finding was that deletion of *GAL6* resulted in an increase in the expression of the galactose-regulated gene, indicating that it is a negative regulator of the GAL regulon. However, the functional significance of this observation is not yet clear.

References

Fried M, Crothers DM (1981) Equilibra and kinetics of lac repressor operator interactions by polyacrylamide gel electrophoresis of protein DNA complexes. Nucl Acids Res 9:6506–6525

Garner M, Revzin A (1981) A gel electrophoresis method for quantifying the binding of protein-to-specific DNA regions. Application to the components of the *E. coli* lactose operon system. Nucl Acids Res 9:3047–3060

Goldberg RF (1974) Autogenous expression of gene expression. Science 183:810–816

Igarashi M, Segawa T, Nogi Y, Suzuki Y, Fukasawa T (1987) Autogenous regulation of the *Saccharomyces cerevisiae* regulatory gene GAL80. Mol Genet 207:273–279

Nogi Y, Fukasawa T (1984) Nucleotide sequence of the yeast regulatory gene GAL80. Nucl Acid Res 12:9287–9298

Nogi Y, Shimada H, Matsuzuki Y, Hashimoto H, Fukasawa T (1984) Regulation of expression of the galactose gene cluster in *Saccharomyces cerevisiae*. Mol Genet 195:29–34

Shimada H, Fukasawa T (1985) Controlled transcription of the yeast regulatory gene GAL80. Gene 39:1–9

Torchia TE, Hamilton RW, Cano CL, Hopper JE (1984) Disruption of regulatory gene GAL80 in *Saccharomyces cerevisiae*: effects on carbon-controlled regulation of the galactose/melibiose pathway genes. Mol Cell Biol 4:1521–1527

Yun S, Hiraoka Y, Nishizava M, Takio K, Titani K, Nogi Y, Fukasawa T (1991) Purification and characterization of the yeast-negative regulatory protein Gal80. J Biol Chem 266:693–697

Zheng W, Eric Xu H, Johnston S (1997) The cystine-peptidase bleomycin hydrolase is a member of the galactose regulon in yeast. J Biol Chem 272:30350–30355

5.5 Isolation of *GAL3*: The Signal Transducer

5.5.1 Introduction

We had discussed that the *gal3* mutants exhibit a delayed growth in response to galactose. Later, it was demonstrated that even transcription of *GAL10* and *MEL1* was delayed in *gal3* mutant as compared to the wild-type (Fig. 5.5.1). Mutations that confer constitutive phenotype, such as deletion of *GAL80*, render the signal-transduction function of *GAL3* irrelevant. That is, a recessive mutation in *GAL80* is epistatic to *GAL3* defect (see sect. 4.1). These observations suggested that in wild-type cells, *GAL3* function is proximal or at the top of the hierarchy in the signal-transduction cascade leading to transcriptional induction. The catalytic model that Gal3p catalyzes the conversion of galactose to a derivative, which in turn abolishes the Gal80p function did not receive experimental support. How Gal3p in conjunction with galactose transmits the signal to inactivate Gal80p, the repressor, remained unclear. Understanding the molecular mechanisms of how galactose transduces the signal had to wait until *GAL3* was cloned and sequenced.

5.5.2 Cloning of GAL3

MEL1 (encodes α-galactosidase) expression is also under the control of the *GAL* genetic switch. A *gal1gal3* strain does not induce α-galactosidase and therefore cannot grow on melibiose. This strain was transformed with the yeast genomic library to uracil prototrophy. Of these, transformants capable of growing on melibiose as the sole carbon source were expected to contain recombinant plasmids bearing the genomic fragment corresponding to *GAL3*, as it would induce

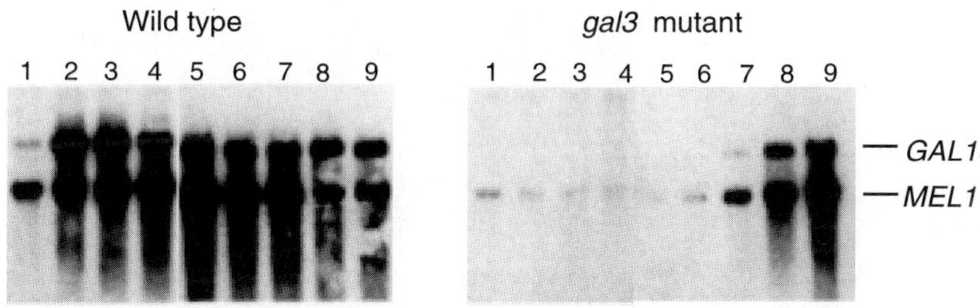

Fig. 5.5.1 Transcriptional response of wild-type and *gal3* mutants to galactose: Total mRNA isolated from wild-type and *gal3* mutant cells grown in glycerol plus galactose was electrophoresed, transferred to nitrocellulose membrane, and hybridized with ^{32}P labeled *GAL10* and *MEL1* probe. *Lanes 1 to 9* represent mRNA samples taken at 5 min, 10 min, 20 min, 2 h, 52 h, 72 h, 108 h and 120 h, for wild-type and 5 min, 10 min, 20 min, 2 h, 42 h, 68 h, 94 h and 114 h for *gal3* mutants (adapted with permission from Torchia and Hopper 1986)

α-galactosidase. Recombinant plasmids isolated from yeast transformants were analyzed. Based on subcloning and insertional mutagenesis, the smallest fragment that complemented the *gal3* growth defect was identified (Fig. 5.5.2). Recall that *TRP1* was shown to be closely linked to GAL3 locus (see 3.2.3). In fact, that the GAL3 plasmid could also complement the *trp1* defect for growth on medium lacking tryptophan was used as genetic evidence to show that genomic fragment present in the recombinant plasmid bears *GAL3* gene.

The genomic *GAL3* locus was disrupted by targeting the in vitro constructed disruption allele using a *LEU2* marker (Fig. 5.5.2). Yeast strain bearing the disrupted *GAL3* locus exhibited long-term adaptation similar to the mutant that was originally isolated. This observation established that the long-term adaptation is due to the loss of *GAL3* protein function and is not unique to the original *gal3* mutant allele. This

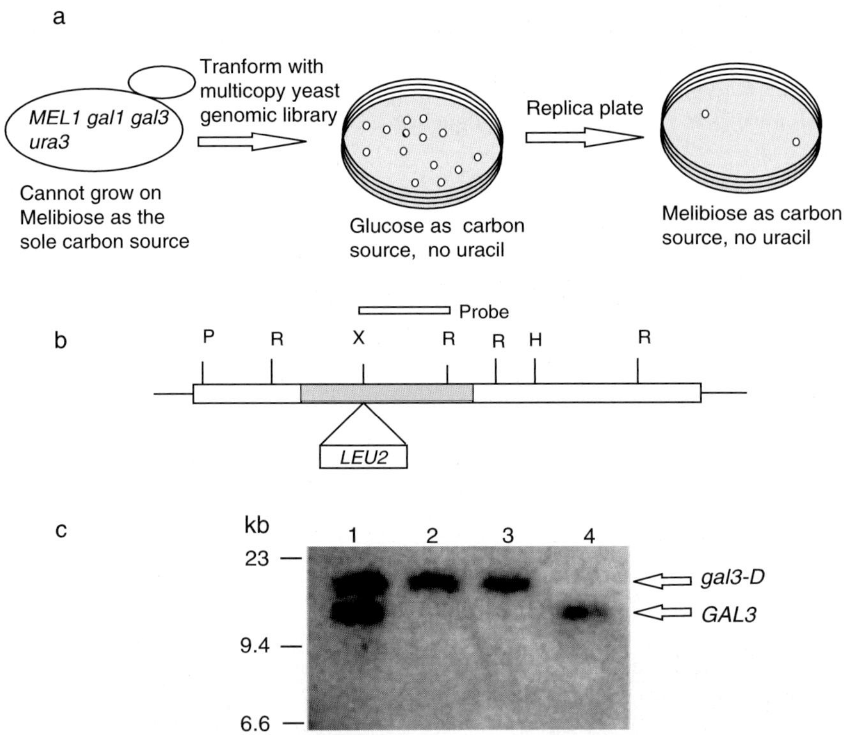

Fig. 5.5.2 Cloning and disruption of *GAL3*. **a** Selection strategy for isolating the *GAL3* complementing recombinant plasmid from the genomic library. **b** The *box* represents the genomic region while the lines on either side represents the plasmid backbone. The *shaded region* corresponding to the functional *GAL3* was delineated by deletion and insertion mutagenesis. *LEU2* gene was inserted at the *Xho*I site to disrupt *GAL3* gene **c** A *leu2* mutant haploid and diploid strain were separately transformed to leucine prototrophy with *Pvu*II and *Hin*DIII digested disrupted plasmid. Southern blot analysis of the *Hin*DIII digested genomic DNA isolated from the putative disrupted diploid (*lane 1*), haploid (*lanes 2 and 3*) and the haploid wild-type. Radioactively labeled 1.3-kb *Eco*R1 and *Hin*DIII fragment was used as the probe (adapted from Torchia and Hopper 1986)

was a significant observation in view of the unusual phenotype exhibited by the *gal3* mutants. The molecular basis of this phenotype is still unclear (see section 8.2.3).

5.5.3 GAL1 and GAL3 are Paralogues

Sequence analysis of the functional region of *GAL3* revealed an open reading frame coding for 520 amino acids. The amino-acid sequence showed striking similarity to the amino-acid sequence of *GAL1* of *Saccharomyces cerevisiae* and *Kluveromyces lactis* as well as other galactokinases. However, galactokinase activity could not be detected in extracts obtained from *gal1* yeast strains harboring multiple copies of *GAL3*, but these extracts did contain Gal3p, which was independently confirmed by Western blot analysis. This observation gave the first clue that *GAL1*, in addition to having galactokinase activity, could also have a signal-transduction function. As expected, over-expression of *GAL1* from a multicopy plasmid suppressed long-term adaptation phenotype. However, over-expression of *E. coli* galactokinase did not suppress the long-term adaptation phenotype, although it could complement that *gal1* defect for growth on galactose. In subsequent experiments, *GAL1* missense mutants C166D and C175Y that lacked kinase activity transduced the signal. These results clearly indicated that the signal-transduction activity is independent of kinase activity pointing out that *GAL1* is a bifunctional protein. *GAL3* appears to have evolved from *GAL1* by duplication followed by loss of kinase activity.

5.5.4 GAL1 is a Degenerate Signal Transducer

That *GAL1* is a true signal transducer is consistent with the observation that *gal-1gal3* is non-inducible. In this strain, both the signal transducers are absent, but it is inconsistent with the non-inducibility of *gal3gal10, gal3gal7, gal3 pgi1, gal3gal5* or *gal3* δ^- strains (note that although these strains do not grow on galactose, the status of *GAL* induction can be monitored by growing these strains except *gal3* δ^- in ethanol and measuring the enzyme activity of any one of the members of the *GAL* family in response to galactose). This observation was not expected, as the above strains contain wild-type copy of *GAL1*. It was surmised that in the above genetic background, *GAL1* is not sufficiently expressed to transduce the signal from galactose. This is because *GAL1* promoter is in turn dependent on the *GAL3* function. If this is true, then the *GAL1* expression from a promoter independent of the galactose control should confer inducibility. Expression of *GAL1* driven by alcohol dehydrogenase promoter in a multicopy plasmid restored the *GAL* gene induction in *gal3* strain lacking the galactose metabolic pathway or mitochondrial function. Overall, the idea that emerged from these studies is that in a *gal3* strain the endogenous *GAL1* acts as a redundant signal transducer, albeit inefficiently. The reason *GAL1* is not expressed in a *gal3gal7* or *gal3* δ^- strain for example is not clear.

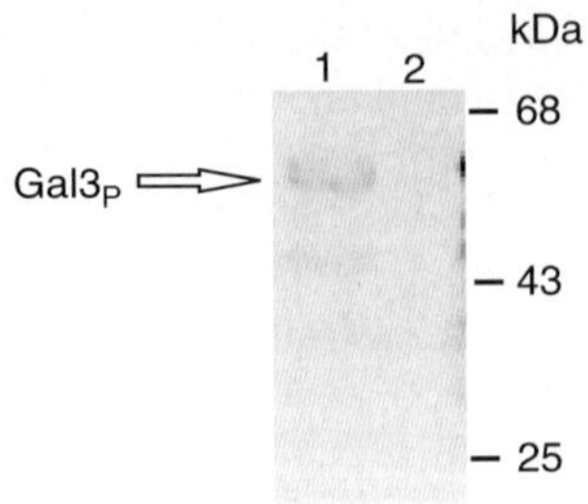

Fig. 5.5.3 Western blot analysis of GAL3 protein. Cell-free extracts obtained form wild-type yeast grown in galactose plus glycerol (*lane 1*), glycerol (*lane 2*), was subjected to SDS electrophoresis, transferred to nitrocellulose filters, and probed with antibodies raised against amino-acid residues starting from I151 to S343 as described below (reproduced with permission from Bhat et al. 1990). A 6-kb DNA fragment encompassing the codons for the above amino acids was isolated as *Bgl*II and *Xho*I fragment cloned in frame with bacterial trpC gene. The fusion protein was over-expressed in *E. coli*. The hybrid protein was isolated by preparative SDS gel electrophoresis and used as the antigen for raising antibodies (adapted with permission from Bhat et al. 1990)

The molecular mechanism of the delay in *GAL1* expression in a *gal3* strain is still an enigma. Secondly, the molecular nature of the galactose-dependent inducing signal is still not clear.

5.5.5 Autogenous Regulation of GAL3 Expression

Northern blot analysis of *GAL3* transcript using the cloned *GAL3* gene as a probe indicated that it is severely repressed in glucose, expressed in glycerol, and induced ten-fold in the presence of galactose. *GAL3* promoter has one UASg element. This induction of *GAL3* was dependent on *GAL4* function. Induction of expression of Gal3p was demonstrated by Western blot analysis using antibodies raised against a part of the Gal3p. In a wild-type cell, *GAL3* protein is detectable in galactose but not glycerol grown cells. Like Gal80p, Gal3p expression is also under the control of *GAL* genetic switch.

5.5.6 An aside on Positional Cloning

Positional cloning refers to the isolation of a gene based on its genetic linkage to a known genetic locus. Since it was already known that *TRP1* and *GAL3* are tightly linked (see sect. 3.2), a recombinant plasmid with a genomic fragment containing *TRP1* locus was expected to contain *GAL3*. Thus, *GAL3* could have been cloned using a positional cloning approach. Yeast centromere was isolated using positional cloning approach. Genetic mapping had indicated that *CDC10* is closely linked to the centromere. A YEp-based recombinant plasmid carrying *CDC10* gene isolated form the genomic library conferred high stability during mitotic and meiotic division, a property expected of centromeric sequences. This indicated that the plasmid in addition to having *CDC10* gene might also have centromeric sequences. Detailed analysis of the recombinant plasmid containing *CDC10* showed that just 130 bp of DNA was all that is required to function as a centromere. At about the same time, a small DNA segment capable of functioning as a telomere in yeast was also isolated. With these elements, it was possible to assemble yeast artificial chromosome, which proved to be instrumental in cloning large DNA pieces that facilitated the genome sequencing effort. Positional cloning is the only approach to clone disease-causing genes that are inherited in a Mendelian fashion for which the biochemical basis is not understood. Once a DNA probe linked to a chromosomal locus of interest is identified, it is possible to isolate the gene in question. Following example illustrates the successful cloning of the cystic fibrosis gene using the above approach.

Neither PON nor cystic fibrosis locus was assigned to any chromosome, although it was known that they are linked (Sect. 3.2). This information was of limited value in identifying the cystic fibrosis gene, as the DNA corresponding to the PON locus was not available. In 1988, a large number of DNA markers were isolated and characterized for mapping purposes using restriction-length fragment polymorphism (see section 5.5.7). Most of these markers showed no linkage to cystic fibrosis, thus excluding 40% of the genome from containing the cystic fibrosis gene. Eventually investigators were successful in identifying a DNA marker linked to cystic fibrosis (Fig. 5.5.4) A total of 43 CF families each with both parents living and at least two or more affected children were screened using the marker for recombinational analysis. The Lod score between the DNA marker and CF was found to be 4.13 at $\theta=0.15$. If this marker is linked to CF and if PON is linked to

Box 5.5.1 *Gene ontology*

Homology represents the relationship of allelic chromosomal segments. It is commonly used to signify common evolutionary origin within or between species. Homology represents a yes-or-no situation. That is, the two elements under comparison are either evolutionarily related or they are not. Homology should not be used as a synonym for similarity. Paralogy represents similarity between non-allelic chromosomal segments. For example *GAL1* and *GAL3* are non-allelic and are paralogues. Paralogy indicates close evolutionary relationship, which may or may not have predated speciation. Xenologues are paralogues arising not out of gene duplication, but by horizontal gene transfer. Orthology indicates close similarity between chromosomal segments or DNA sequences between species. For example, the *GAL1* of yeast is an ortholog of *gal*K of *E. coli* or GK1 of humans. Analogue are two evolutionarily unrelated protein but perform the same functions. Synologues are evolutionarily two distinct sequences

Table 5.5.1 Linkage relationship between the polymorphic marker PON and CF (data obtained with permission from Tsui et al. 1985)

Loci	Families	0.05	0.10	0.15	0.20	0.30	0.40	θ	Z
CF-marker	39	1.67	3.63	3.95	3.62	2.18	1.38	0.14	3.96
CF-PON	11	5.01	4.78	4.28	3.66	2.25	1.51	0.05	5.01

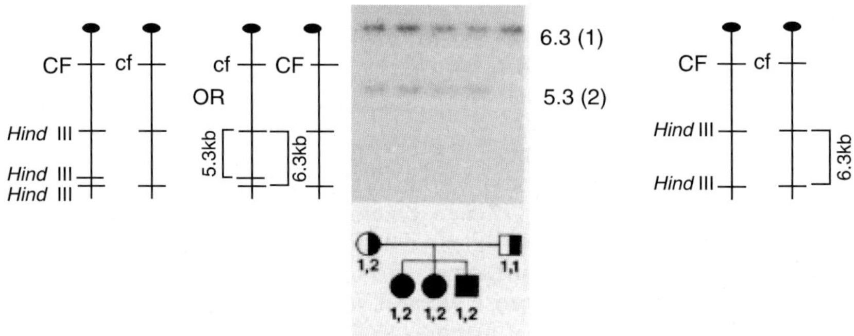

Fig. 5.5.4 Linkage analysis using a restriction fragment length polymorphic marker in a representative family. *Half-filled symbols* represent carriers for cystic fibrosis and *full shaded symbols* are individuals with cystic fibrosis. Total genomic DNA digested with *Hin*DIII was electrophoresed on an agarose gel, transferred to nitrocellulose membrane, and probed with ^{32}P labeled probe. This probe was previously shown to detect polymorphic loci with two alleles with a polymorphism information content of 0.3. The mother is heterozygous for the marker locus as she has both 6.3 (allele 1) and 5.3-kb (allele 2) fragments while the father is homozygous with both the alleles being 6.3 (CF and cf indicate wild type and mutant alleles respectively). However, in the mother, the phase of the linkage between the marker locus and the cystic fibrosis locus is not known. (data obtained with permission from Tsui et al. 1985)

CF, then the marker should show linkage to PON. Serum samples from the same families were analyzed and the families were typed using the above probe. The Lod score between marker and the PON locus was found to be 5.01 at $\theta = 0.05$. These data placed the PON locus to be in between CF and the marker. In an independent study based on human mouse hybrid cell analysis, the polymorphic DNA marker locus was shown to be on chromosome 7. These observations initiated a massive hunt for cystic fibrosis on chromosome 7. Using the DNA marker, overlapping DNA fragments from the human genomic library were isolated moving progressively closer to the cystic fibrosis locus. Finally, DNA markers that were in linkage disequilibrium with the cystic fibrosis locus were isolated. Linkage disequilibrium between the marker and the disease locus is an indication that no recombination between these two loci has occurred. Of the 4 ORFs located within the suspected region, a potential ORF corresponding to cystic fibrosis was identified by locating a inframe deletion of phenylalanine at position 508, only in cystic fibrosis patents but not in a normal individual. The gene was found to code for a protein (cystic fibrosis transmembrane regulator) of 1,480 amino acids and shown to be required for the conductance of chloride ion across the membrane.

5.5.7 *Restriction Fragment Length Polymorphism*

A large number of substitution, deletion, or addition mutations are not associated with visible phenotype. However, these mutations in DNA sequences are revealed as differences in the size of fragments after restriction digestion followed by Southern hybridization. These differences are inherited stably. Such differences became powerful signatures because any arbitrary stretch of DNA could be expected to function as a potential genetic marker. The idea of using RFLP as a marker system for mapping genes was proposed in a genetics retreat (held at Alta, USA; in 1978) sponsored by the University of Utah. Individuals heterozygous for such mutations can be identified by digesting DNA samples with relevant restriction endonucleases and identifying specific restriction fragment whose lengths are characteristic of the allele, so-called RFLPs. In the example discussed in Fig. 5.5.4, because of a mutation at one of the *HinD*III site, two alleles of fragment length of 6.3 and 5.3 can be identified. Any individual can be of the genotype 6.3/6.3 or 5.3/5.3, or 6.3/5.3, genetic constitution (RFLP markers are co-dominant). Since the mother was heterozygous for this allele, recombinants between this and the cystic fibrosis locus could be identified. By 1996, a large array of DNA markers based on variable number of tandem repeats and microsatellites were identified and their position on the chromosomes determined. This data was instrumental in mapping the human genome.

Polymorphism information content. In Fig. 5.5.4, the father was homozygous with respect to the marker locus and therefore not informative. Even if both the parents were to be heterozygous for the same set of alleles the marker would have been noninformative. The marker's usefulness for linkage analysis depends on the number of alleles and their gene frequencies. That is, any family used for analysis should have heterozygosity at the maker locus consisting of a different pair of alleles. Heterozygosity is the measure of polymorphisms $H = 1 - \Sigma p_i^2$ where H is the heterozygosity and p is the frequency of the ith allele. For example, a frequency of three alleles of a given locus is 0.28,. 06, and.66, which yields a heterozygosity of 0.48. This means that almost half of the individuals are expected to be heterozygous at that locus. Heterozygosity is largest when the alleles are equally frequent, $pi = pj = 1/n$. For example, for a heterozygosity of 0.9, there should be ten alleles with each allele having a frequency of 0.1. Another measure of heterozygosity is the polymorphism information content $(PIC) = 1 - \Sigma p_i^2 - \Sigma\Sigma 2p_i^2 2p_j^2$ Here, the third term takes half the matings of similar heterozygotes. For example, for a marker with two alleles with equal frequency the heterozygosity will be 0.5 but the PIC is only 0.375.

References

Bajwa W, Torchia TE, Hopper JE (1988) Yeast regulatory gene GAL3: carbon regulation, UASg elements common with GAL1, 2, 7, 10, 80 and MEL1; encoded protein is strikingly similar to yeast and *E. coli* galactokinases. Mol Cell Biol 8:3439–3447

Bhat PJ, Oh D, Hopper JE (1990) Analysis of the Gal3 signal-transduction pathway activating Gal4 protein-dependent transcription in *Saccharomyces cerevisiae*. Genetics 125:281–291

Bhat PJ, Hopper JE (1991) The mechanism of inducer formation in Gal3 mutants of the yeast galactose system is independent of normal galactose metabolism and mitochondrial respiratory function. Genetics 128:133–139

Botstein D, White RL, Skolnick M, Davis RW (1980) Construction of a genetic linkage map in man using restriction fragment length polymorphism. Am J Hum Genet 32:314–331

Collins FS (1992) Cystic fibrosis: molecular biology and therapeutic implications. Science 256:774–779

Knowlton et al (1985) A polymorphic DNA marker linked to cystic fibrosis is located on chromosome 7. Nature 318:380–382

Leppert MF (1990) Gene mapping and other tools for discovery. Epilepsia 31 (Suppl 3): S11–S18

Mayers J, Walker-Jonah A, Hollenberg CP (1991) Galactokinase encoded by GAL1 is a bifunctional protein required for induction of the *GAL* genes in *Kluyveromyces lactis* and is able to suppress the *gal3* phenotype in *Saccharomyces cerevisiae*. Mol Cell Biol 11:5454–5461

Mount DW (2001) Bioinformatics: Sequence and genome analysis. Cold Spring Harbor Laboratory Press. New York

Ott J (1991) Analysis of human genetic linkage. The John Hopkins University Press, Baltimore

Schmiegelow et al (1986) Linkage between the loci for cystic fibrosis and paraoxonase. Clin Genet 29:374–377

Torchia TE, Hopper JE (1986) Genetic and molecular analysis of the GAL3 gene in the expression of the galactose/ melibiose regulon of *Saccharomyces cerevisiae*. Genetics 113:229–246

Tsui Lap-Chee et al (1985) Cystic fibrosis locus defined by a genetically linked polymorphic DNA marker. Science 230:1054–1057

White R, Laiouei J (1988) Chromosomal mapping with DNA markers. Scientific American 258:20–28

Wilmut I et al. (1997) Viable offspring derived from fetal and adult mammalian cells. Nature 385:810–813

Chapter 6
Signal Transduction Revisited

6.1 Revised Model of Signal Transduction

6.1.1 Introduction

The signal from galactose is transmitted to Gal80p-Gal4p transcriptional switch through Gal3p/Gal1p signaling system. Based on the genetic experiments discussed previously, it appeared unlikely that signaling is due to a derivative of galactose catalyzed by Gal3p. Attempts to dissociate Gal4p-Gal80p complex in presence of galactose derivatives were unsuccessful, but it was clear that at least galactokinase activity is not required for signaling. Therefore, an idea that necessitates the presence of galactose for induction but does not warrant that the signal transducer to be an enzyme was considered.

6.1.2 Protein–Protein Interaction Model

According to this, Gal3 protein exists in an equilibrium between active and inactive forms. Galactose shifts the equilibrium in favor of the active form to a level above the threshold required for induction. This implied that overproduction of Gal3p could increase the concentration of active Gal3p above a threshold required for induction and should lead to constitutive expression of *GAL* genes. As predicted, over-expression of Gal3p from multiple-copy plasmid in the absence of galactose induced *GAL* enzymes 40% of the fully inducible wild-type level. Based on these results, it was inferred that galactose triggers Gal3p or Gal1p to inactivate Gal80p by direct interaction (Fig. 6.1.1).

The model predicts that it is possible to demonstrate (a) interaction of galactose with Gal3p, (b) galactose-dependent conformational change in Gal3p, (c) direct interaction between Gal3p and Gal80p, and (d) isolation of constitutive mutants of Gal3p that transduce the signal in the absence of galactose. While the last two predictions have come true, direct experimental evidence for galactose-binding and galactose-dependent conformational change has not been forthcoming. In this

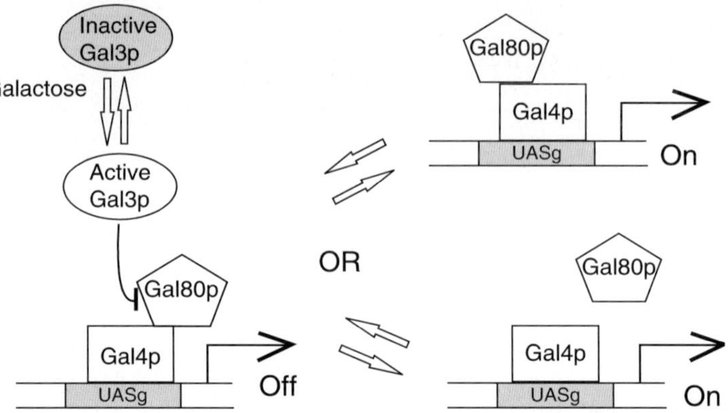

Fig. 6.1.1 Illustration of protein–protein interaction model. In a wild-type cell, inactive and active Gal3p exist in equilibrium. Galactose transforms the inactive Gal3p to an active form, which abolishes the repressor function of Gal80p by direct protein–protein interaction. During this process, either Gal80p protein is dissociated from the Gal4p-Gal80p complex or the interaction is altered such that Gal4p is free to activate the transcription (adapted with permission from Bhat & Hopper 1992)

section, I shall discuss the experiments that demonstrated that Gal3p and Gal80p indeed interact.

6.1.3 Testing the Predictions of the Protein–Protein Interaction Model

Physical interaction between Gal3p and Gal80p was demonstrated by independent laboratories using co-immuno-precipitation or pull down assays either from whole cell extract or using pure Gal3p and GAL80p. As predicted by the model, Gal3p and GAL80p interact even in the absence of galactose, but it is weak, in that the interaction can be abolished by including 50 mM NaCl in the reaction mixture. This weak interaction is stabilized by ATP in the presence of galactose but not other monosaccharides (Fig. 6.1.2). While galactose cannot be substituted by any other hexose as a ligand, ATP can be substituted by other nucleotides. However, hydrolysis of nucleotides is not required for the interaction. Recall that *GAL3* is a paralogue of *GAL1* and does not posses kinase activity. Therefore, it is not surprising that it uses ATP as a ligand and not as an energy source. Missense mutants of Gal3p that are unable to transduce the signal did not bind Gal80p in vitro. Similarly, dominant Gal80 mutant proteins such as Gal80^{s-0} are unable to interact with Gal3p in vitro. These observations are consistent with the model that inactivation of Gal80p occurs by direct interaction with the Gal3p in the presence of galactose and ATP.

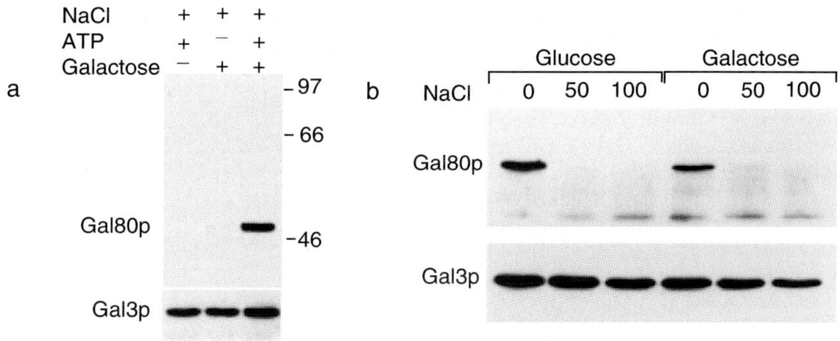

Fig. 6.1.2 Demonstration of GAL3-GAL80 protein interaction. **a** Whole cell extract prepared from yeast cells over-expressing Gal80p/HA-Gal3p, grown in non-fermentable carbon source was immuno-precipitated using HA antibody and analyzed by western blot analysis using Gal80p (upper panel) or HA antibody (lower panel). The concentration of NaCl, ATP and galactose in the reaction mixture was 100, 2 and 1mM respectively. **b** Whole cell extract prepared from cells over-expressing Gal80p/HA-Gal3p grown in glucose or galactose as indicated, imuno-precipitated and analyzed as before. The reaction mixture contained indicated concentration of NaCl in the absence of galactose and ATP (data obtained with permission from Yano et al. 1996)

While the above model appears to be compatible with many features of the genetic switch, it is not clear how the signal is relayed to Gal80p-Gal4p complex that resides on UASg elements. Using an in vitro transcription system it was demonstrated that the transcription of the reporter gene was dependent on the presence of purified GAL4 (1–93+768–881) protein. Transcription was inhibited if pure Gal80p was included in the reaction mixture. The presence of a pure constitutive Gal3p and galactose could overcome the repression conferred by Gal80p. Employing a gel retardation experiment, a tripartite complex consisting of pure Gal4p, Gal80p, constitutive Gal3p and the UASg was demonstrated (Fig. 6.1.3). The idea that emerged from these studies is that Gal3p interacts with Gal80p-Gal4p complex and activates transcription without causing dissociation of Gal80p from Gal4p. That Gal80p and Gal4p do not dissociate upon induction was also supported by dihybrid analysis (section 8.4.4). Interestingly, in an independent study, it was demonstrated that during induction, Gal80p translocates to a second site on Gal4p. All these experiments meant that Gal3p needs to move into the nucleus to transduce the signal.

6.1.4 *Recent Analysis of Signal Transduction*

A detailed study of the sub-cellular localization using indirect immunofluorescence and cell fractionation studies indicated that Gal3p is exclusively located in the cytoplasm, whereas Gal80p is distributed both in cytoplasm as well as the nucleus.

Fig. 6.1.3 Demonstration of a tripartite complex. Radioactively labeled consensus UASg was incubated with purified GAL80 or GAL4 (GAL4 1-93+768-881) or GAL3^c proteins or in combination as indicated and subjected to gel mobility assay. Free DNA (i), DNA-Gal4p (ii), DNA-Gal4p-Gal80p (iii) and DNA-Gal4p-Gal80p-Gal3 complex (iv). (reproduced with permission from Platt and Reece 1998)

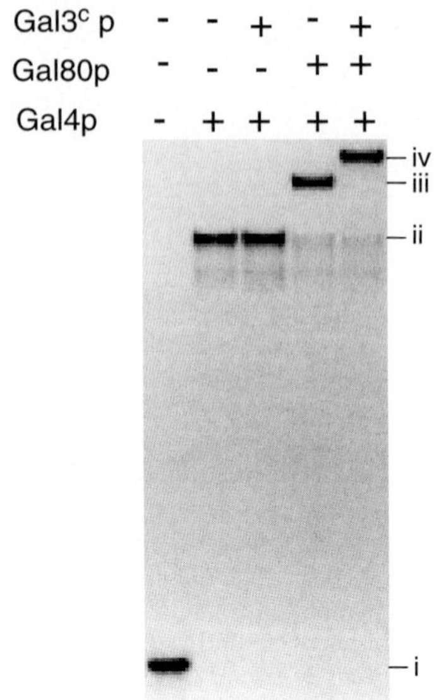

It was further demonstrated that if Gal80p protein is forced to get concentrated into the nucleus by fusing it to ectopic nuclear localization signal sequences, *GAL* gene induction is severely impaired, indicating that shuttling of Gal80p between cytoplasm and the nucleus is an intrinsic feature of the signaling mechanism. These observations invoked the possibility of Gal3p/Gal1p trapping Gal80p in the cytoplasm as a means of relieving Gal4p from the Gal80p repression upon induction.

In a follow-up study, independent lines of evidences demonstrated that Gal3p transduces the signal by trapping Gal80p in the cytoplasm. In one of these experiments, a *gal3gal1* strain (both the signaling pathways are abolished) was transformed with a plasmid containing wild-type *GAL3*, or *GAL3* ORF fused with myristylation signal (Myr-Gal3p) that directs the protein to the plasma membrane. That this tagged protein is localized to the plasma membrane was independently confirmed by observing the distribution of wild-type Gal3p, Myr-gal3p fused to green fluorescent protein using fluorescence microscopy. In these transformants, *GAL* gene induction was monitored by estimating α-galactosidase activity (Fig. 6.1.4a). It is clear that despite Gal3p being anchored on the plasma membrane α-galactosidase expression occurred, indicating that Gal3p need not enter the nucleus for induction.

In another experiment, in vivo association of Gal80p, Gal3p and Gal4p with UASg element under inducing and non-inducing conditions was monitored by chromatin immunoprecipitation assay (see Box 6.1.1). It was reasoned that if these proteins remain associated even upon induction, Gal3p antibody should pull down the UASg. Similarly, if Gal80p and Gal4p remain bound upon induction then antibody against Gal80p should pull down the UASg element. Under induced conditions, neither Gal80p nor Gal3p antibodies pulled down the UASg element, but Gal4

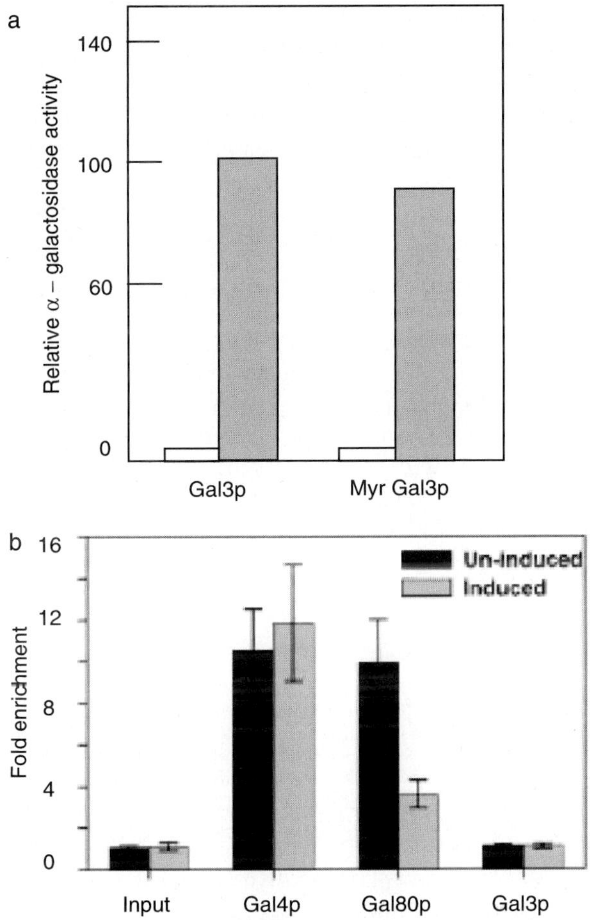

Fig. 6.1.4 Demonstration of Gal3p, Gal80p interaction in the cytoplasm. **a** expression of α-galactosidase in a *gal3gal1* strain expressing wild-type or derivative of Gal3p fused to myristylation signal (Myr-Gal3p). *Open* and the *shaded bar* represent the relative activity under non-inducing and inducing condition, respectively. **b** Relative enrichment of UASg obtained using chromatin immunoprecipitation assay with antibodies against Gal4p, Gal80p and Gal3p under both induced and uninduced conditions (reproduced with permission from Peng and Hopper 2002)

antibodies did (Fig. 6.1.4b). These results argue that upon induction, Gal80p is unlikely to be associated with Gal4p. Overall, these experiments suggest that most likely Gal3p traps Gal80p in the cytoplasm there by reducing the effective concentration of Gal80p in the nucleus resulting in the transcriptional activation.

Analysis of signal transduction by FRET analysis: The mechanism of Gal3p signaling was studied using fluorescence resonance energy transfer (FRET) analysis (see Box 6.1.2). Both Gal4p and Gal80p were expressed as fusion proteins of enhanced

Box 6.1.1 *Chromatin immuno-precipitation assay*

This is used to detect the in vivo interaction of protein-DNA or higher-order protein–protein DNA complexes. Here, the immuno-precipitation is carried out with specific antibodies after cross linking the protein and the chromosomal DNA. This is followed by shearing the DNA and reversing the crosslink followed by identifying the specific DNA fragment or quantitating the difference between the experimental and the control using PCR. In the example discussed above, this technique was exploited to determine whether Gal3p and Gal80p remain bound even after induction. This technique has also been used to identify genes that are under the control of Gal4p transcription. In this case, the DNA fragments obtained from immuno-precipitated and the pre-immun-serum treated samples were amplified and labeled and with Cyanin5 and Cyanin 3, respectively, by ligation mediated polymerase chain reaction to interrogate a microarray of all the known yeast intergenic sequences. In principle, this can also be used to identify unknown proteins associated with the protein under consideration.

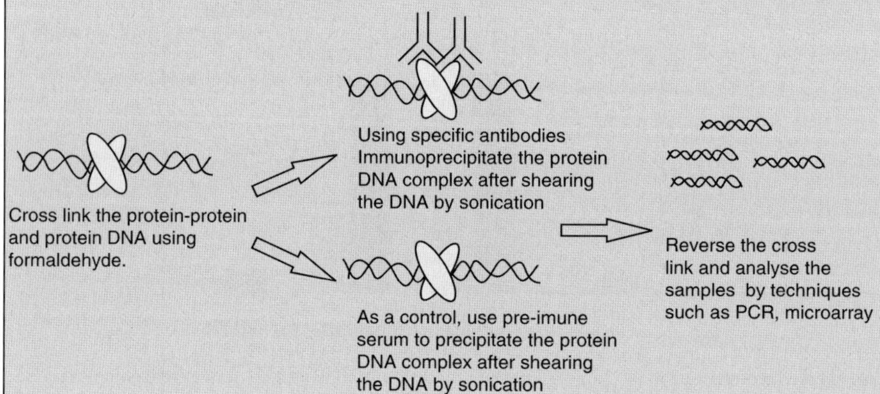

Fig. 6.1.5 Schematic representation of chromatin immuno precipitation assay

cyan fluorescent protein (GAL4-ECFP) and enhanced yellow fluorescent protein (GAL80-EYFP), respectively. In un-induced conditions, yeast cells expressing these fusion proteins remain associated and, as expected, give a FRET signal. Upon induction, if GAL80-EYFP is sequestered in the cytoplasm, then the FRET signal is expected to disappear. However, it was observed that the FRET signal persisted even after induction, indicating that these two proteins remain together or close enough to give a FRET signal.

Results discussed thus far propose two mechanistically different pathways for transducing the signal from galactose to Gal4p transcriptional activator. One model proposes a reorientation of the interaction between Gal4p and Gal80p that allows

Box 6.1.2 *Fluorescent probes for in vivo studies*

Sub-cellular localization of macromolecules in intact cells is determined by direct or indirect labeling techniques using fluorescent dyes. For example, DAPI, a fluorescent dye, intercalates with DNA and fluoresces upon excitation, which is monitored using a fluorescent microscope. Localization of proteins is tracked by first allowing them to interact with the antibody and then tracking the antigen-antibody complex using the fluorescent labeled second antibody. This is called the indirect immunofluorescence technique. In fact, the sub-cellular distribution of Gal3p and Gal80p was demonstrated using this technique.

Green fluorescent protein (GFP) of jelly fish *Aequorea victoria* has proved to be a powerful, non-invasive, sensitive in vivo probe to monitor various intracellular processes such as, gene expression, sub-cellular distribution, protein–protein interaction, etc. GFP has been mutated to yield variants with enhanced fluorescence intensity and overlapping spectral properties, making it an ideal system to track the protein fate in vivo. The protein under consideration is tagged with any of these variants using recombinant DNA technology. In most of the cases, the hybrid proteins retain the biological activity. These fusion proteins are then expressed from the native promoters to maintain the same level of expression. Interaction between any two proteins can be determined by expressing the fusion proteins with overlapping spectral properties followed by FRET analysis. Sub-cellular localization is followed by fluorescence microscopy. Transcriptional induction in single cells is followed by FACS analysis.

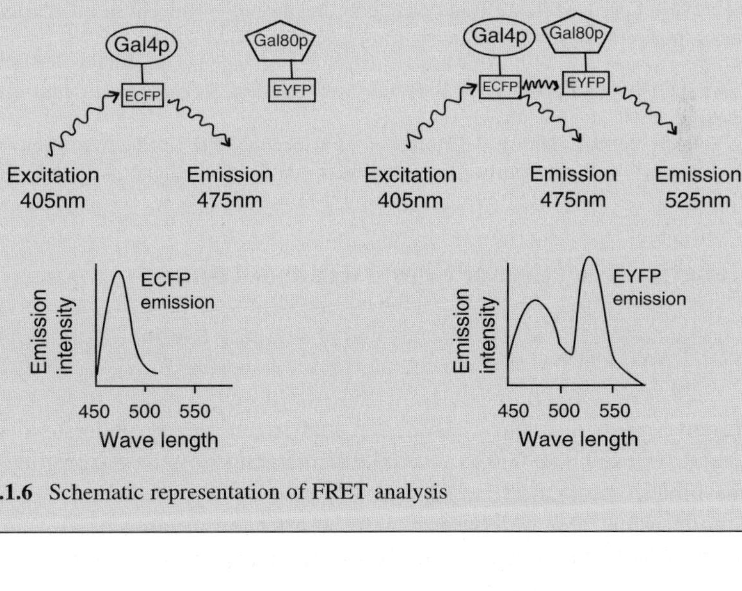

Fig. 6.1.6 Schematic representation of FRET analysis

Gal4p to activate transcription. The other, cytoplasmic trapping of Gal80p, envisions a reduction in the effective concentration of Gal80p in the nucleus, thereby allowing Gal4p to activate transcription. Which of these two possibilities are true remains to be determined.

References

Bhaumik SR, Raha T, Aiello DP, Green MR (2004) In vivo target of a transcriptional activator revealed by fluorescence resonance energy transfer. Genes Dev 18:333–343

Bhat PJ, Hopper JE (1992) Overproduction of the GAL1 or GAL3 protein causes galactose-independent activation of the GAL4 protein. Evidence for a novel model of induction for the yeast *GAL*/MEL regulon. Mol Cell Biol 12:2701–2707

Kuo M, Allis CD (1999) In vivo cross linking and immunoprecipitation for studying dynamic protein: DNA associations in a chromatin environment. Methods 19:425–433

Peng G, Hopper JE (2000) Evidence for Gal3p's cytoplasmic location and Gal80p's dual cytoplasmic nuclear location implicates new mechanisms for controlling Gal4p activity in *Saccharomyces cerevisiae*. Mol Cell Biol 20:5140–5148

Peng G, Hopper JE (2002) Gene activation by interaction of an inhibitor with a cytoplasmic signaling protein. Proc Nat Acad Sci USA 99:8548–8553

Platt A, Reece RJ (1998) The yeast galactose genetic switch is mediated by the formation of a Gal4p-Gal80p-Gal3p complex. EMBO J 17:4086–4091

Sil AK, Alam S, Xin P, Ma L, Morgan M, Lebo CM, Woods MP, Hopper JE (1999) The Gal3p-Gal80p-Gal4p transcription switch of yeast: Gal3p destabilizes the Gal80p-Gal4p complex in response to galactose and ATP. Mol Cell Biol 19:7828–7840

Suzuki-Fujimoto T, Fukuma M, Yano K, Sakurai H, Vonika A, Johnston SA, Fukasawa T (1996) Analysis of the galactose signal-transduction pathway in *Saccharomyces cerevisiae*: interaction between Gal3p and Gal80p. Mol Cell Biol 16:2504–2508

Yang T, Sinai P, Green G, Kitts PA, Chen Y, Lybarget L, Chevranak R, Patterson GH, Piston DW, Kain SR (1998) Improved fluorescence and dual color detection with enhanced blue and green variants of the green fluorescent protein. J Biol Chem 273:8212–8216

Yano Ken-Icji, Fukasawa T (1997) Galactose-dependent reversible interaction of Gal3p with Gal80p in the induction pathway of Gal4-activated genes of *Saccharomyces cerevisiae* Proc Nat Acad Sci USA 94:1721–1726

Zenke FT, Vollenbroich RE, Meyer VJ, Hollenberg CP, Breunig KD (1996) Activation of Gal4p by galactose-dependent interaction of galactokinase and Gal80p. Science 272:1662–1665

6.2 Genetic Dissection of Signal Transduction

6.2.1 Introduction

An important aspect of signal transduction that is not clearly understood is how galactose and ATP stabilize Gal3p and Gal80p interaction. While the original model of protein–protein interaction between Gal3p and Gal80p predicts that galactose induces a conformational change in Gal3p, neither the binding of galactose nor the presumed conformational change has been detected. The structure-activity relationship of Gal3p and Gal80p has been extensively studied by mutational analysis with the hope of gaining an understanding of Gal3p-Gal80p interaction. Further, Gal1p is crystallized and its three dimensional structure has been determined at a resolution of 2^{0}A, but the precise mechanism of how galactose and ATP transduce the signal remains elusive.

6.2.2 Mutational Analysis of GAL3

Gal3p is a member of *g*alactose, *h*omoserine, *m*evalonate and *p*hosphomevalonate kinase super family. Gal1p and Gal3p of *Saccharomyces (GHMP) cerevisiae* as well as Gal1p of *Kluveromyces lactis* are considerably larger as compared to galactokinases of human and bacterial origin (Fig. 6.2.1). Gal3p and Gal1p show 73% identity and 92% similarity. *GAL3* arose from *GAL1* by gene duplication followed by loss of galactokinase activity. *GAL3* lacks the conserved serine and alanine residues present in the galactokinases of *S. cerevisiae* and *K. lactis*. Introduction of these two residues in *GAL3* re-instated galactokinase activity albeit to a very low level. Deletion of the first 28 N terminal and last nine C-terminal amino acids of Gal3p did not inactivate the protein, but the stability of internal deletion mutants of *GAL3* is severely compromised. These results implied that Gal3p is a globular protein sensitive to mutations.

Three dimensional structure of yeast Gal1p with α D-galactose and Mg- adenosine (β, γ-imido) triphosphate as ligands has been determined at 2.4A^0 resolution. The protein exhibits a bilobal structure, the active site being wedged between the N- and C-terminal domains. The two lobes are connected by a "hinge". A homology model of Gal3p derived based on the three-dimensional structure of Gal1p of *S. cerevisiae* (Fig. 6.2.2) provides a convenient starting point to rationalize the behavior of constitutive mutants of Gal3p.

Substitution mutations in Gal3p that lead to constitutive induction of *GAL* regulon are spread throughout the primary sequence. For example, F237Y, D368V and S509D/L confer constitutive induction of *GAL* genes. Of these, F237 and S509 are located in the hinge region while D368 is remote with respect to the above two mutations (Fig. 6.2.2). To further evaluate the mechanistic basis of the constitutive activity of these alleles, D68S Gal3 allele was isolated. This substitution suppressed the constitutive induction function of F237Y, and 509P/L. However, D68S did not suppress the D368V constitutive activity. D68 is located in the putative hinge region. Based on these studies it has been suggested that galactose binding induces a conformational change brought about by the hinge movement. The constitutive mutations such as F237Y and S509D probably mimic the structure induced by galactose. This is in contrast to the constitutive activity of D368V, which is located in an insertion region present only in *GAL1, GAL3* and *GAL1* of *K. lactis*, but not in galactokinases of other organisms and appears to be Gal80p interacting domain. Consistent with this, the D68S substitution does not suppress the D368V constitutive function. The implications of these studies is that the hinge movement

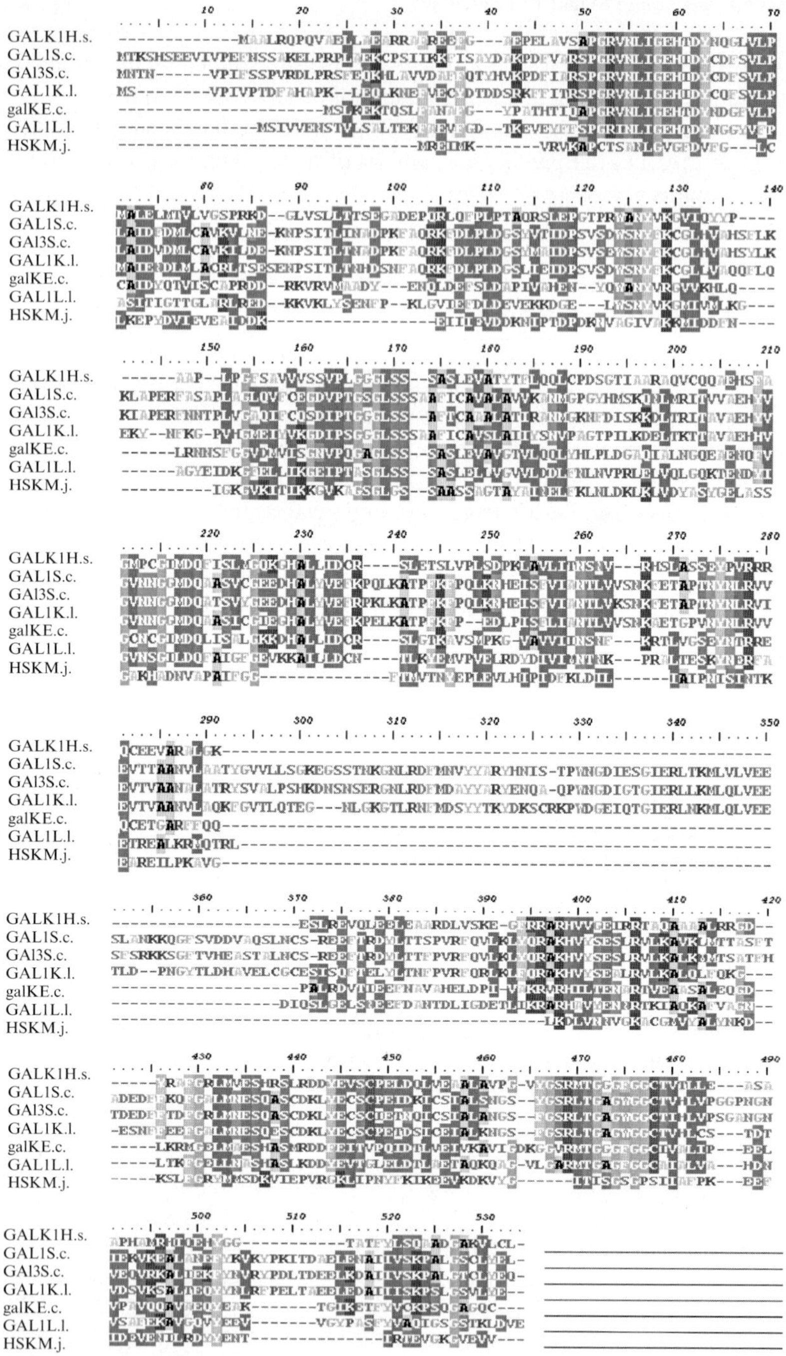

Fig. 6.2.1 Multiple sequence alignment of galactokinases. Galactokinase of *Homo sapiens* (GALK1 H.s) *Saccharomyces cerevisiae* (GAL1 S.c), *Kluyveromyces lactis* (GAL1 K.l), *Lactobacillus lactis* (galK L.c) and GAL3 of *Saccharomyces cerevisiae* (GAL3 S.c), homoserine kinase of *Methanococcus jannaschii* (HSK M.j)

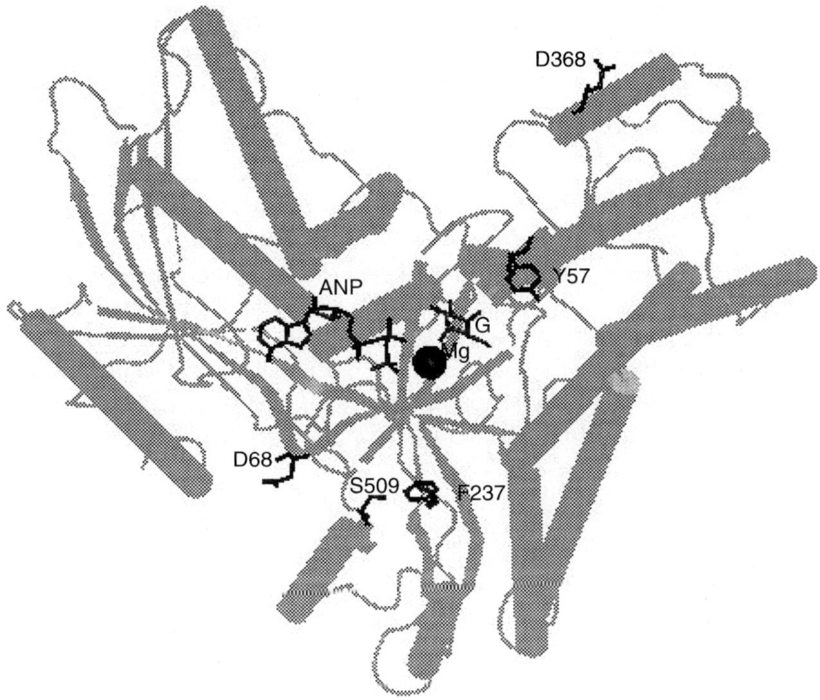

Fig. 6.2.2 Homology model of Gal3p based on the structure of its paralogue Gal1p of *Saccharomyces cerevisiae. ANP* is adenyl imido triphosphate, *G* is galactose and *Mg* is magnesium. Some relevant amino-acid residues are indicated (coordinates for the homology model were provided by Reece 2005)

could be a crucial step in the galactose-induced conformational change leading to signal transduction.

A Gal3p derivative consisting of Y57W substitution is defective in responding to low external galactose but its response to high galactose is comparable to that of the wild-type. A unique property of this mutant protein, however, is that upon over-expression, the constitutive induction activity is significantly less compared to the constitutive activity obtained by over-expressing wild-type Gal3p. That this lowered constitutive induction activity is not due to the lowered protein is ruled out based on the Western blot analysis. These observations led to the idea that Y57W mutant protein appears to have an intrinsic propensity to exist in inactive conformation as compared to the wild-type protein.

While the mutational analysis seems to suggest that galactose binding induces an allosteric change leading to the recognition of Gal80p, there is no direct experimental demonstration of this phenomenon. An alternate possibility is that Gal3p and Gal80p associate and dissociate continually and galactose only reduces the off rate of Gal3-Gal80 dissociation, thus leading to the transient sequestration of Gal80p in the cytoplasm. That is, two distinct active and inactive states of Gal3p may not exist. Instead, Gal3p exists in active and inactive form probably through a breathing motion across the hinge even in the absence of galactose. This might explain the failure to capture the transition from inactive to active form thought to be induced by galactose.

6.2.3 Mutational Analysis of GAL80

Gal80p occupies a pivotal position in the working of the *GAL* genetic switch. Upon purification, it behaves as a monomeric protein, but in vivo it exists as a homodimer with a dimerization constant of the order of 10^{-10} M. The reason for not detecting the dimmers of Gal80p during purification appears to be due to its high off rate. In addition to the above intramolecular interaction, Gal80p interacts with Gal4p as well as Gal3 protein. All these activities are tightly packed into a protein of relatively small size consisting of 435 amino acids.

A genetic screen to identify missense mutations that knock off its repressor function was set up to establish structure activity relationship. Of the 21 such mutants analyzed, none had retained Gal4p, Gal3p or dimerization function. This data suggested that the Gal80p is too sensitive to mutational perturbations. This is unlike Gal4p in that its domains can be separated without the loss of function of other domain.

Table 6.2.1 Properties of mutant derivatives of Gal80p. Dimerization defect between mutant and mutant as well as mutant and wild-type is separately indicated

Substitution	Defect in GAL4 binding	Defect in Gal3 binding	Defect in dimerization	
			Mut X Mut	WT X Mut
V144D	Yes	Yes	–	+
G152D	Yes	No	+	+
G183S	Yes	No	+	+
R189G	Yes	No	+	+
D260G	Yes	Yes	–	–
H261R	Yes	No	+	+
V275E	Yes	Yes	–	+
G282S	Yes	No	+	+
G301R	No	Yes	ND	ND
A309T	Yes	No	+	+
G310D, E	Yes	No	+	+
L319P	Yes	No	+	+
G323R	No	Yes	ND	ND
E351K	No	Yes	ND	ND
V352E	No	Yes	+	+
Y366V	No	Yes	ND	ND
Y369N	Yes	Yes	–	+
D404G	Yes	Yes	–	+
L406P	Yes	Yes	–	+

Box 6.2.1 *Convergent evolution*

Three distinct families of sugar kinases, hexokinase (e.g., glycerol kinase, ribulokinase, xylulokinase) ribokinase (e.g., 6-phosphofructokinase, fructokinase, 6-phosphotagatokinae) and galactokinases (e.g., mevalonate kinase, homoserine kinase, phosphomevalonate kinase) have been identified. It is observed that within each family there is a divergence of substrate specificity. For example, in the galactokinase family, the substrate specificity varies from sugars to intermediates of amino-acid metabolism. It is known that three families of sugar kinases have different three-dimensional structures and the conserved sequence pattern are strikingly different, pointing out that the chemically equivalent reactions of phosphorylation of structurally related molecules have evolved from three distinct structural frameworks. What is more striking is that even within the family, the substrate specificities appear to have evolved independently. For example, galactose and mevalonate are quite different. What this means is that active sites can be adapted onto similar three-dimensional folds to bring in different specificities and it is the three-dimensional folds that are consistent in evolution.

In an independent study, mutants that specifically lacked Gal3p-binding activity or Gal4p-binding activity were sought by employing a specific selection strategy involving dihybrid assay (section 8.4.4). It turned out that most of the mutants identified to be defective for Gal4p binding were capable of Gal3p binding indicating that Gal3p and Gal4p-binding domains are independent. In contrast to this, most of the mutants selected for impairment of Gal3p binding had defect in Gal80p dimerization. Based on this, it was suggested that Gal3p probably prevents Gal80p from dimerization by directly binding with monomer units. In fact, experimentally, it has been demonstrated that Gal3p and Gal80p interact with a stoichiometry of 1:1. This led to the idea that probably under in vivo conditions dimerization of Gal80p is an intrinsic part of the signal-transduction mechanism.

References

Bork P, Sander C, Valencia A (1993) Convergent evolution of similar enzymatic function on different protein folds: the hexokinase, ribokinase and galactokinase family of sugar kinases. Protein Sci 2:31–40

Diep C, Gang P, Beweley M, Pilauri V, Ropson I, Hopper JE (2006) Intragenic suppression of Gal3[c] interaction with Gal80 in the *Saccharomyces cerevisiae GAL* gene switch. Genetics 172:77–87

Lakshninarasimhan A, Bhat PJ (2005) Replacement of a conserved tyrosine by tryptophan in Gal3p of *Saccharomyces cerevisiae* reduces constitutive activity: implication for signal transduction in the *GAL* regulon. Mol Genet Genom 274:384–393

Melcher K (2005) Mutational hypersensitivity of a gene regulatory protein: *Saccharomyces cerevisiae* Gal80p. Genetics 171:469–476

Menezes RA, Amuel C, Engels R, Gengenbacher U, Labahan J, Hollenberg C (2003) Sites for interaction between Gal80p and Gal1p in *Kluveromyces lactis*: structural model of galactokinase based on homology to the GHMP protein family. J Mol Biol 333:479–492

Nogi Y, Fukasawa T (1989) Functional domains of a negative regulatory protein GAL80 of *Saccharomyces cerevisiae*. Mol Cell Biol 9:3009–3017

Pilauri V, Beweley M, Diep Q, Hopper JE (2005) Gal80 dimerisation and the yeast *GAL* gene switch. Genetics 169:1903–1914

Platt A, Ross HC, Hankin S, Reece R (2000) The insertion of two amino acids into a transcriptional inducer converts it into a galactokinase. Proc Nat Acad Sci USA 97:3154–3159

Thoden JB, Sellick CA, Timson DJ, Reece RJ, Holden H (2005) Molecular structure of *Saccharomyces cerevisiae* Gal1p, a functional galactokinase and transcriptional inducer. J Biol Chem 280:36905–36911

Timson DJ, Ross HC, Reece RJ (2002) Gal3p and Gal1p interact with the transcriptional repressor Gal80pto forma complex of 1:1 stoichiometry. Biochem J 363:515–520

Chapter 7
Versatile Galactose Genetic Switch

7.1 Transcription Activation

7.1.1 Introduction

In eucaryotes, three distinct species of RNA polymerases (RNA Pol) have been identified while prokaryotes and archaea have only one RNA polymerase. In eukaryotes, RNA PolI, transcribes the large rRNA precursor that is processed leading to the production of 28 S, 18 S and 5.8 S rRNAs. RNA PolII, transcribes heterogeneous nuclear RNAs (hnRNAs) and small nuclear RNAs (sn RNAS). hnRNAs are processed to mature mRNA, which requires snRNAs. RNA PolIII transcribes 5 S rRNA and tRNAs. The identity of these three RNA polymerases has been confirmed by molecular genetic and biochemical analysis. Although these enzymes share a common property of transcribing DNA, they lack the ability to identify the transcription initiation sites for which they depend on additional proteins. As expected, the promoters of genes transcribed by these RNA polymerases also have unique features.

Of the three polymerases, PolII has attracted considerable attention owing to the fact that its activity is highly regulated. The fundamental question is: How does the transcriptional machinery direct the RNA PolII to accurately transcribe the diverse set of genes depending upon the physiological need? In general, promoters served by the RNA PolII have a common architecture consisting of core elements required for the promoter function and for the assembly and orientation of the pre-initiation complex. The most significant structural elements are the TATA sequences located 25 nucleotides upstream of the transcription initiation site. The transcription initiation site is generally pyrimidine-rich. In addition, the promoters contain regulatory sequences that govern the expression status by interacting with specific transcriptional regulators. *GAL* promoter is an archetypical PolII promoter that is turned OFF by glucose and ON by galactose. We shall focus on how Gal4p, the DNA-binding transcriptional activator, activates PolII to transcribe *GAL* genes in response to the inducer.

P.J. Bhat, *Galactose Regulon of Yeast.*

7.1.2 RNA Polymerase II

RNAPolII holoenzyme is an unusually large complex consisting of the core subunits RPB 1 to 12. The three-dimensional structure of yeast RNA polymerase has been solved at 2.3A˙ resolution. This core polymerase is associated with six complexes of general transcription factors, TFIIA. TFIIB, TFIID, TFIIE, TFIIF and TFIIH. That these factors are required for proper transcription became clear when it was shown that in the presence of these factors, purified PolII could accurately transcribe the native adenovirus DNA as the template. These fractions were separated from nuclear extracts by eluting from ion-exchange columns at different salt concentrations. These are collectively called "general transcription factors". In addition to the above, other coactivator complexes such as SRB (suppressors of RNA polymerase B)mediator complex, SRB10 CDK (cyclin-dependent kinase) complex, Swi/Snf chromatin remodeling complex, SAGA (*S*PT, *A*DA, *G*CN5, *A*ceyl transferase) the chromatin modifying complex (Table 7.1.1) are also required. Thus, the RNA polymerase holoenzyme is a complex of more than 70 polypeptides. While the core subunit is required for the transcription of all the genes driven by PolII promoters, other complexes of coactivators may be required for the transcription of only few genes. For example, in yeast, the SRB10/CDK complex is required for 3% of the total genes transcribed by PolII. This goes to suggest that these additional polypeptides are not required for the transcription per se, but serve regulatory functions.

The biochemical activity and the three-dimensional structure of many of these proteins have been reported. For example, TFIID is a complex of TATA-binding protein and ten other associated factors. The TATA-binding protein is highly evolutionarily conserved and the first component of TFIID to be characterized. The most important function of TFIID is to recognize the core promoter and possibly interact with gene-specific transcriptional activators. The high-resolution structure of TBP bound to DNA clearly demonstrates that the binding induces a bend in the DNA. Due to this, the DNA is partially unwound, and is thought to assist the subsequent

Table 7.1.1 RNA polymerase II core and its associated complexes with a representative component from each group

Name of the complex	Components	Subunits
RNA polymerase II	RBP1	12
Mediator/SRB complex	GAL11	22
SRB10CDK complex	SRB10	4
General transcription factors		
TFIIA	TOA1	2
TFIIB	SAU7	1
TFIID	TBP	13
TFIIE	RAP74 (?)	2
TFIIF	TFG1	3
TFIIH	RAD53	9
Nucleosome modifiers (SAGA complex)	GCN5	13
Swi/Snf Complex (Histone remodellers)	SWI2	11

assembly of other protein factors. TFIIA has two subunits while TFIIB has only a single subunit. TFIIA stabilize the binding between TFIID and promoters TFIIB serves as a linker between TFIIF and PolII. TFIIF is composed of two subunits. TFIIH phosphorylates the carboxy terminal and has helicase activity that is essential for transcription.

The mediator complex consists of 20 subunits. It was observed early on that over-expressing transcriptional activators without their DNA-binding domain severely interfered with the transcription of genes that are normally not served by the over-expressed truncated transcriptional activator. This phenomenon was referred to as "squelching". It implied that the transcriptional activator titrated the transcriptional factors, thus making them unavailable for normal transcription. In in vitro transcription experiments, increasing the concentration of one activator inhibited the activation of another activator apparently by the same mechanisms, but further addition of general transcription factors did not overcome this defect, indicating that these TAFs are not the targets for squelching. A protein fraction that could overcome squelching was isolated and was found to have 20 subunits. This was referred to as a "mediator" as it did not increase the basal expression but increased the expression brought about by the transcriptional activator.

SRB proteins, the members of the mediator complex, were identified in the following genetic screen. Wild-type *GAL4* cannot function in yeast strains bearing a RNA PolII lacking the C-terminal domain. Mutations that suppressed this phenotype were sought. Several suppressor mutations in genes named SRBs for "suppressors of RNAPolB" were identified. These gene products were later analyzed for their role in transcription activation. Similarly, the chromatin modeling and remodfier complex consists of many proteins.

7.1.3 Transcriptional Activation by Recruitment

It is quite clear that under any given physiological condition, only a fraction of the genes required to fulfill the needs of the cells are transcribed. To ensure this, the cell is endowed with a limited supply of core RNA polymerase and depending upon the signal, it is activated or recruited to transcribe the genes required under a particular physiological condition. Unlike typical metabolic enzymes, PolII does not have unique substrate specificity, nor does its activation follow the typical mechanisms observed for large majority of metabolic enzymes. As discussed before, the assembly of the pre-initiation complex is therefore highly regulated, involving many functions encoded by a diverse set of proteins.

Transcription initiation involves the assembly of general transcription factors on the core promoter to form the pre-initiation complex. This is followed by the isomerization of the polymerase-DNA complex from a closed to an open form in which the DNA strands are locally separated so that polymerase can access the template strand. The activator does not appear to function as the mediator of this isomerization, rather, it appears that transcriptional activators, recruit the RNA PolII holoenzyme to the promoter. Enzyme recruitment as a mechanism of activation of enzymes is not restricted to transcription alone. There are other processes where multiprotein complexes such as proteolysis, splicing, wherein enzymes are recruited to perform their dedicated role.

How does Gal4p recruit the PolII holoenzyme? Many of the properties of transcription activating sequences are shared by topogenic sequences, peptides that mark the proteins for transport in and out of the nucleus. When tested in vitro, the acidic activation region of Gal4p showed interaction with several members of transcriptional machinery, making it difficult to identify the true target of Gal4p in vivo. Based on a large body of genetic and biochemical studies, the members of the mediator complex, the components of TFIID and SAGS complex have been implicated to be the targets of the transcriptional activators, but the functional significance of such interactions is still being worked out.

Given the participation of a large number of polypeptides, two distinct possibilities were originally envisioned to explain the formation of the pre-initiation complex. Based on the in vitro transcription experiments in the presence of protein fractions obtained by chromatographic separation, it was speculated that the core RNAPolII and other transcription factors are recruited sequentially in a stepwise fashion to the promoter. The second possibility is that holoenzyme is recruited en block after the TATA-binding protein identifies the core promoter element. This is based on the observation that PolII holoenzyme could be purified as a preassembled holoenzyme consisting of many coactivators such as mediator complex. Of the two mechanisms, evidence is more in favor of the recruitment model.

Genetic evidence in support of recruitment of the holoenzyme as a mechanism of transcription activation came from the activator bypass experiment conducted in yeast. A yeast strain bearing GAL4 1–147+AH (a derivative of GAL4 consisting of N-terminal 147 residues tuned to ampiphathic & helix) weakly expresses *GAL* genes and is unable to grow on galactose as the sole carbon source. This strain was mutagenized and mutant capable of growing on galactose better than the parent strain was isolated at a frequency of 10^{-5}. The mutant gene responsible for overcoming the defective transcriptional activity of GAL4 1–147+AH was isolated form the genomic DNA library of the mutant strain. The gene was found to be the same as *GAL11* and the mutant allele was referred to as *GAL11*P (P for potentiation). *GAL11* was originally identified in a genetic screen as a recessive mutation that reduces the *GAL* gene expression by 30–40%. Further analysis of *GAL11*P indicated that a substitution of isoleucine by asparagine at position 342 conferred this ability to activate the *GAL* genes by the weak Gal4 1–140+AH. *GAL11* was later shown to be required for the transcription of many other genes and is a member of the mediator complex.

In vitro experiments confirmed that Gal11Pp interacts with amino acids 58–97 of GAL4 (1–100). An independent in vivo experiment was carried out to further confirm the interaction. A fusion protein consisting of DNA-binding domain of LexA fused to GAL4 (58–97) was expressed in a wild-type or GAL11P mutant strain bearing lexA:β-galactosidase reporter cassette. Strain with GAL11P mutant allele but not with wild-type *GAL11* expressed β-galactosidase. In a follow-up experiment it was shown that a strain bearing *GAL11* fused to LexA DNA-binding domain activated the β-galactosidase expression. These experiments suggest that Gal4p recruits the holoenzyme probably by interacting with some members of the complexes.

The in vivo targets of Gal4p interaction have been identified by employing various techniques. Using in vitro approaches, the transcriptional activation domain has been shown to be interacting with TFIIB, TFIIH, TBP and its associated factors of the TFIID complex. More specifically, the 34 amino-acid core activation domain of Gal4p interacts with Gal80p, Sug1p, Ada2p, and TBP. Of these, TBP interaction with the activation domain of Gal4p has been extensively studied. Based on these results, it has been suggested that interaction between TBP and Gal4p activation domain is the rate determining step in the transcriptional activation by Gal4p.

Box 7.1.1 *Higher-order structure of DNA*

The length-to-breadth ratio of a human chromosome is ten million to one with a length of ~2 m. If this has to be accommodated within a small volume of nucleus without getting tangled, it has to be properly folded. In eucaryotes, the double-stranded DNA is wrapped around an octamer of basic proteins histones H2A, H2B, H3, and H4 (two molecules of each). The DNA length required to wrap around is 140 bp. The linker DNA between the nucleosomes is 55 bp. The whole chromosome appears like a string with beads. Histone H1 is bound on the outer aspect of nucleosome where the DNA enters and leaves the core particle. The next higher-order structure is the solenoid loop of 30 nm, consisting of 12 nucleosomes. This unit contains up to 50 kB of DNA. Six of these loops form a rosette; 30 of these rosettes make one coil and ten coils make a chromosome.

In this scenario, how does one visualize the activation of gene transcription by DNA-binding transcription activators such as GAL4? It is clear that the nucleosomes are, in general, repressors of transcription and active promoters are generally devoid of nucleosomes. In some active genes there are precisely positioned nucleosomes in the control regions. There is growing experimental evidence to indicate that nucleosomes are removed from promoters during the course of transcriptional activation. Two mechanisms were considered. In one, the hisotone octamer was thought to slide away from the promoter region while the other considered that the dislodging of the histone octamer from the DNA. Elegant experiments have supported the second model. In some cases, histone acetylation has been shown to stabilize the nucleosomal stability, thereby preventing the transcription activation. In yeast, genes such as Swi/Snf bring about remodeling of nucleosomes to facilitate the transcription activation.It has been shown from in vitro experiments that as the RNA polymerase traverses through the nucleosomal DNA, it displaces the nucleosomes behind, thus clearing its path for transcription.

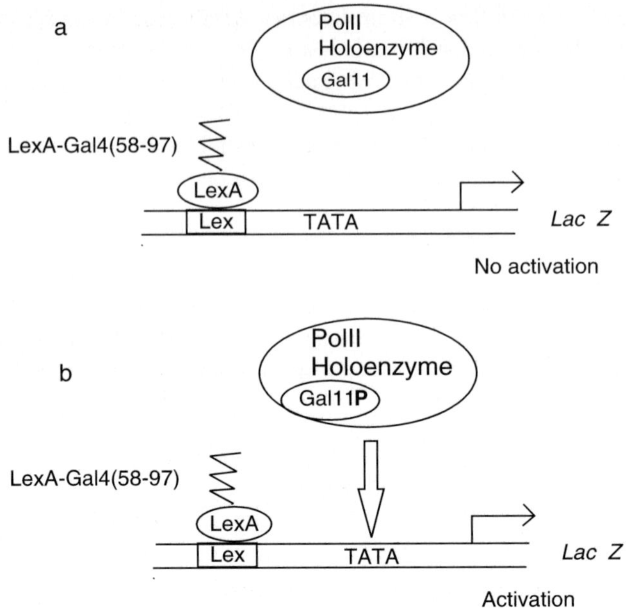

Fig. 7.1.1 Transcriptional activation by Gal11P protein. **a** A yeast strain with an integrated copy of *lac*Z expression cassette with *lex*A operator locus and a *lex*A-Gal4 (55–97) fusion protein. This strain does not activate the expression of *lac*Z. **b** Strain bearing GAL11P, expresses the *lacZ* gene indicating that it can interact with the *lex*A-Gal4 (55–97) fusion protein

More recently, using chromatin immunoprecipitation assays (see Chap. 6.1) it was shown that upon induction by galactose, mediator complex associates with UASg and not with the core promoter elements, implying that Gal4p recruits RNA PolII through the mediator. This was unexpected, as the mediator was thought to interact with the core promoter elements as it formed a part of RNA PolII complex, and this association was independent of RNA PolII, further suggesting that the recruitment probably occurs through the mediator complex. However, using a similar approach, another study demonstrated that during induction, SAGA, mediator complex, RNA PolII followed by other TAFs complexes bind in a sequential manner with a time gap of 2–3 min. According to this, a three-stage recruitment process has been suggested. Of these, only PolII moves into the gene almost immediately after its recruitment on to the core promoter for transcription.

Recently, FRET analysis has been used to detect protein–protein interaction during Gal4p mediated transcriptional activation. A panel of strains co-expressing Gal4p-ECFP (enhanced cyan fluorescence protein) and one of the 14 SAGA members fused to EYFP (enhanced yellow fluorescence protein) were generated. These strains were induced and subjected to FRET analysis. Of these, a FRET signal was observed only in strain co-expressing Gal4p-ECFP and Tra1-EYFP, a subunit of

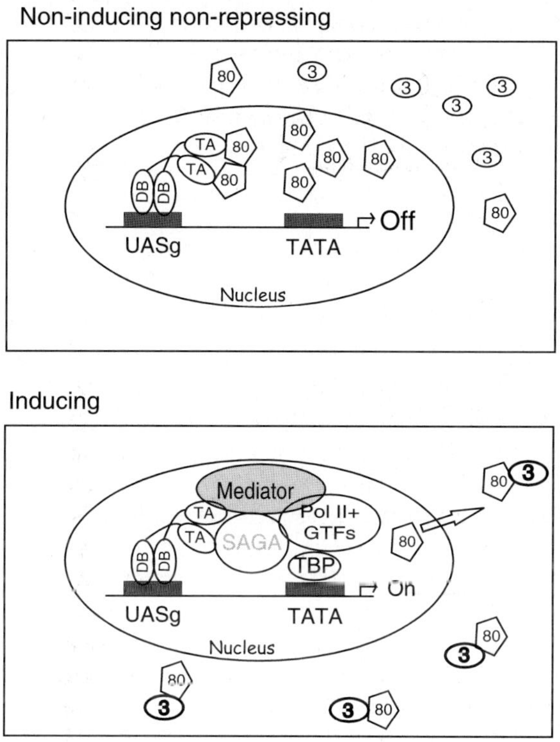

Fig. 7.1.2 Schematic representation of activation of transcription by Gal4p

SAGA complex, indicating that Gal4p most likely recruit SAGA complex through Tra1 subunit. This interaction initiates the recruitment of mediator followed by the core Pol II and TAFs, which form the pre-initiation complex at the core promoter elements.

It is clear from the above that multiprotein complex machinery transcribes thousands of genes. Components of this machinery are remarkably conserved across species. Gal4p tethered to DNA is capable of activating transcription in plants, insects, and mammalian cells, demonstrating the functional conservation. In eucaryotes, the DNA exists as a complex structure consisting of nucleosomes and is generally refractory to transcription to ensure that a positive signal is required for the activation of otherwise repressed genes. Based on extensive studies it has been deduced that nucleosomes disappear from a transcriptionally active promoter. Here, the mechanism appears to be that the nucleosomes are removed by disassembly rather than move away from the promoter and SWI/SNF complex has been implicated in this process. RNA PolII complex is established on nucleosome free promoters. It has been observed that Gal4p-binding sites present in *GAL1-10*, *GAL7* and *GAL80* promoters are nucleosome-free regions.

References

Boeger H, Bushnell DA, Davis R, Griesenbeck J, Lorch Y, Strattan JS, Westover KD, Kornberg RD (2005) Structural basis of eukaryotic transcription. FEBS Lett 579:899–903

Bryant GO, Ptashne M (2003) Independent recruitment in vivo by Gal4 of two complexes required for transcription. Mol Cell 11:1301–1309

Frank CP, Holstege et al (1998) Dissecting the regulatory circuitry of a eukaryotic genome. Cell 95:717–728

Himmelfarb H, Pearlberg JJ, Last DH, Ptashne M (1990) GAL11P: a yeast mutation that potentiates the effect of the weak GAL4 derived activators. Cell 63:1299–1309

Kornberg RD (2005) Mediator and the mechanism of transcriptional activation. Trends Biochem Sci 30:235–239

Kornberg RD, Lorch Y (1999) Twenty-five years of the nucleosome: fundamental particle of the eukaryote chromosome. Cell 98:285–294

Nogi Y, Fukasawa T (1983) A novel mutation that affects utilization of galactose in *Saccharomyces cerevisiae*. Curr Genet 195:115–120

Ptashne M (2003) Regulated recruitment and co-operativity in the design of biological regulatory systems. Phil Trans R Soc Lond A 361:1223–1234

Reece RJ (2000) Molecular basis of nutrient-controlled gene expression in *Saccharomyces cerevisiae*. CMLS Cell Mol Life Sci 57:1161–1171

Reece RJ, Platt A (1997) Signalling activation and repression of RNA polymerase II transcription in yeast Bioassays 19:1001–1009

Sellick CA, Reece RJ (2005) Eucaryotic transcription factors as direct nutrient sensors. Trends Biochem Sci 30:405–412

Traven A, Jelicic B, Sopta M (2006) Yeast GAL4: a transcriptional paradigm revisited. EMBO Rep 7:496–499

Thomas MC, Miang CC (2006) The general transcription machinery and general co-factors. Crit Rev Biochem Mol Biol 41:105–178

Tsonis PA (2003) Anatomy of gene regulation. Three-dimensional structural analysis. Cambridge University Press, Cambridge

7.2 Glucose Repression

7.2.1 Introduction

Glucose is the most abundant and preferred source of carbon and energy. Microorganisms in particular have evolved complex genetic regulatory mechanisms to shut off the expression of proteins required for the metabolic pathways that are unnecessary when glucose is used as the source of carbon and energy. Negative regulation by glucose probably allows microorganisms to sustain with the minimum metabolic machinery required to propagate itself. This phenomenon is referred to as carbon catabolite repression or simply glucose repression. In yeast, glucose repression is exerted at multiple levels such as transcription, protein and/or mRNA stability, translation post-translational modification, etc. Another feature of glucose repression is that some genes are repressed at glucose concentrations as low as 0.01%, while repressions of other gene may require a five-fold higher concentration of glucose. Obviously, yeast has evolved mechanisms to sense and turn on different glucose-signaling pathways depending upon the context.

Broadly, two classes of genes are identified based on the nature of expression of genes upon glucose exhaustion. For example, invertase, an enzyme involved in the utilization of sucrose, is expressed upon glucose withdrawal. In other words, the expression of invertase is under negative regulation by glucose. In contrast, withdrawing glucose alone is not sufficient to induce the *GAL* genes expression. It also needs a positive signal from galactose. Thus galactose genetic switch responds to glucose and galactose in a reciprocal fashion. With respect to *GAL* switch glucose inhibition occurs at three levels. One, at the level of galactose transport by reducing the availability of Gal2p, second by reducing the level of the signal transducer, that is at the level of *GAL3/GAL1* and third by far the most dominant mechanism is repressing the transcription of *GAL4*. Here I shall focus on the role of *GAL80* and *MIG1* in glucose repression.

7.2.2 MIG1 Protein is a DNA-Binding Transcriptional Repressor

Mig1p is a nucleocytoplasmic shuttle protein that belongs to the C_2H_2 zinc finger, GC box-binding protein and is a homologue of Wilms tumor protein. In the absence of glucose, Mig1p is phosphorylated at multiple sites and largely remains in the cytoplasm. The available evidence suggests that Mig1p is phosphorylated by heterotrimeric SNF1 kinase complex and dephosphorylated by a complex of phosphatases encoded by *GLC7* and *REG1*. SNF1 protein kinase is a homologue of the adenosine monophosphate-activated protein kinase of higher eucaryotes. Based on this, it has been proposed that SNF1 kinase might respond to glucose in a similar fashion (Fig. 7.2.1). Experimental evidence suggests that low AMP:ATP ratio favors the inactivation of Snf1p. Under conditions of low AMP:ATP ratio, Mig1p is under-phosphorylated and enters the nucleus, binds to its cognate binding sites present in the promoters of *GAL1, GAL3* and *GAL4*. Repression of *GAL1* and *GAL3*, in addition to *GAL4* probably ensures that even the pre-existing Gal4p is unable to activate the transcription. *MIG1* recruits other general repressor proteins such as TUP/SSN6 to repress the transcription of these genes.

7.2.3 Combined Role of GAL80 and MIG1 Proteins in Glucose Repression

It has been reported quite early on that *GAL80* is involved in mediating glucose repression. *GAL80* is the only member of the *GAL* gene family that is not repressed by glucose. This physiological attribute is expected as the normal function of Gal80p is to inhibit Gal4p function. It has been demonstrated that glucose repression of *GAL1* is abolished in a *gal80* strain in which either *MIG1*-binding sites or *MIG1* is deleted. This indicates that glucose repression does not affect Gal4p post-translationally. In wild-type cells (in the presence of *MIG1* and *GAL80*) *GAL1* expression is repressed

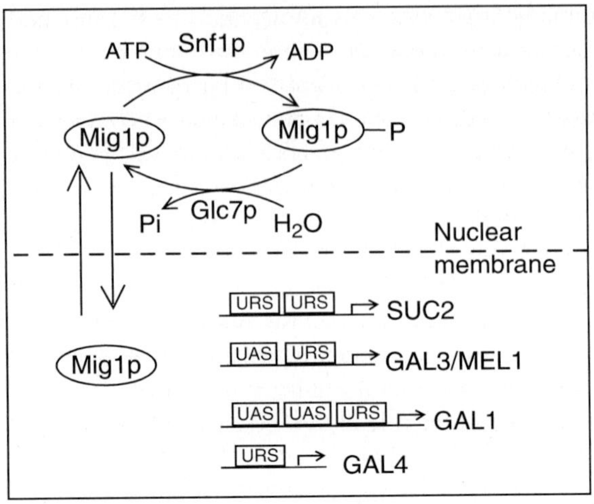

Fig. 7.2.1 Mig[1]P mediates glucose repression. A large number of genes are under the control of MIG1 and for the sake of clarity only few genes are shown. Mig1p binds to the upstream regulatory sequence (URS) elements present on *SUC2, GAL3, MEL1, GAL1* and *GAL4* in response to glucose. *GAL1, GAL3* and *MEL1* expression is also indirectly repressed through *GAL4* repression

Table 7.2.1 Glucose repression in strains with indicated genetic background; values represent the percentage of the mRNA levels expressed in wild-type at 2% galactose in the absence of glucose, indicated by bold (data obtained with permission from Nehlin et al. 1991)

	MIG1GAL80		*mig1GAL80*		*MIG1gal80*		*mig1gal80*	
2% Glucose	+	−	+	−	+	−	+	−
2% Galactose	+	+	+	+	+	+	+	+
expression (%)	1.2	**100**	4.7	100	6.9	90	60	98
Repression (%)	98.8		95.3		93.1		40	

by glucose to an extent of 99% (Table 7.2.1) and *GAL1* gene is repressed by 40% in strains disrupted for both *MIG1* and *GAL80*. However, cells disrupted for either *MIG1* or *GAL80* alone still exhibit glucose repression to the extent of 95%. Thus, *GAL80* and *MIG1* act synergistically to repress the *GAL* gene expression in response to glucose. Even after abolishing glucose repression by inactivating *MIG1* and *GAL80* residual glucose repression still persists. This could be because of post-transcriptional mechanisms such as glucose inactivation. For example, Gal2p has been reported to be inactivated in the presence of glucose.

7.2.4 Binary and Graded Response

In general, transcriptional response to a repressor or an inducer is monitored by estimating the mRNA or enzyme levels as an average of a population of cells. However, in a population, cells could exist in a different state of induction or repression, which will not be apparent if a population average of reporter gene expression is obtained. That cells with a different steady-state expression exist in a population was originally shown in *E. coli* β-galactosidase expression as early as 1957. It is now possible to evaluate the induction or repression state at the single-cell level by monitoring the fluorescence of intrinsically fluorescing proteins whose expression is driven by the promoter under study (see Box 7.2.1).

Using yeast cells with single copy of chromosomally integrated *GAL1* promoter fused to green fluorescence protein, transcriptional status of *GAL1* in response to galactose was measured. The *GAL1* expression increased in a graded fashion over a 500-fold change in galactose concentration. That is, all the cells in the population were responding as a function of galactose concentration. This analysis was then extended to monitor the repression by glucose mediated by *MIG1* and *GAL80*, either individually or in concert. Wild-type and *MIG1* deleted cells (in this strain glucose repression would be through Gal80p) exhibited a binary response when glucose concentration was increased at a fixed concentration of galactose. That is, the population

Box 7.2.1 *Binary and graded response in gene expression*

A fundamental question being actively pursued is whether transcriptional regulation serves as a binary switch between active and inactive states or serves as a rheostat where the response is graded. When the yeast cell population as a whole is experimentally analyzed for transcriptional response, graded response cannot be distinguished from binary or vice versa. For instance, if the induction is say 50% of the maximal at the population level, it is not possible to assert whether (a) all the cells are expressing at 50% level or only 50% of the cells are expressing 100%, resulting in an average induction of 50%. If the former is true, then it means that as the induction signal is increased (for example, increase in galactose concentration), all the cells gradually keep increasing their output; this is called "graded response". If the latter is true, cells are either fully induced or not induced at all. As the induction signal increases, the fraction of cells fully induced keeps increasing; this is called "binary response". New reporter systems have enabled the analysis of transcriptional induction at single cells and have revealed that genes are either maximally expressed or not expressed at all.

Expression at the single-cell level is determined by monitoring the fluorescence of green fluorescent protein. After the induction, fluorescence is monitored in individual cells through flow cytometry. Cell suspension is introduced into a liquid jet and passed individually through a laser beam. As each cell passes through, it emits a flash of fluorescence, which is quantitative. Many thousands of cells can be can be quickly analyzed in this way. In fact, the cells can be sorted into different groups depending on the intensity. If it is graded, the fluorescence intensity keeps changing, but not the cell number. If it is binary, the number of cells keep changing at a given fluorescence intensity.

(continued)

Box 7.2.1 (continued)

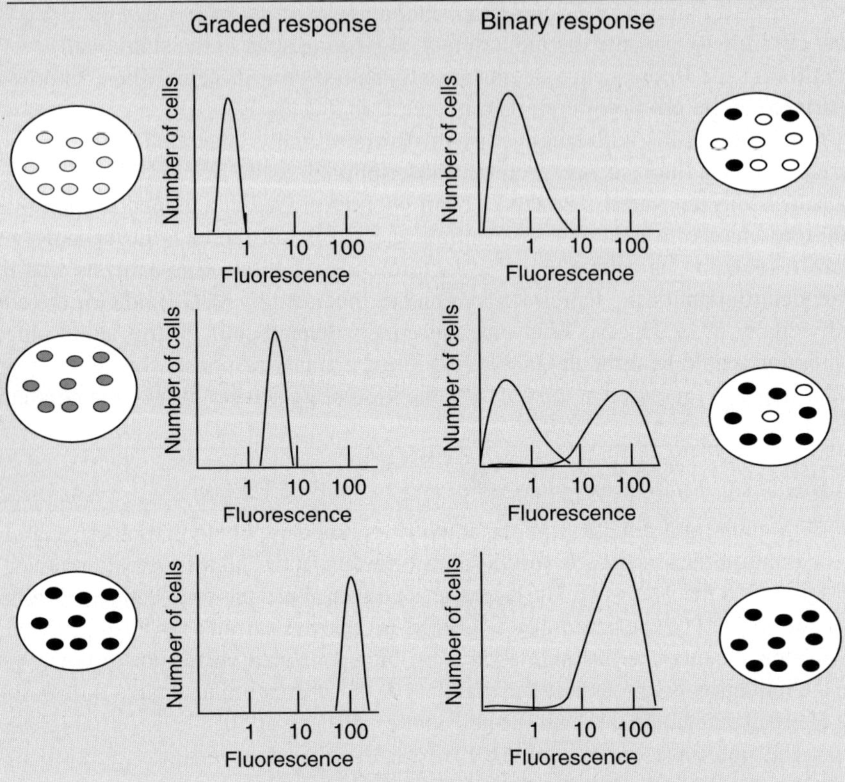

Fig. 7.2.2 Schematic representation of binary and graded response. Each small *circle* in the disc represents a cell and the gene-expression status is indicated by the intensity of the shade; a *light shade* indicates low expression, while a *dark shade* indicates full induction. In graded response, cell population as a whole responds gradually in response to the induction signal. In binary, the fraction of cells fully responding keeps increasing

has a different proportion of cells that are either fully repressed or not repressed at all. The proportion of cells in the population fully repressed increased as a function of glucose concentration. In contrast, in a *gal80* disrupted strain (glucose repression is caused through *MIG1* pathway in *gal80* cells) at a fixed galactose concentration increasing glucose concentration resulted in a graded response. That is, *GAL1* repression in all the cells of the population increased as a function of increasing glucose concentration. These studies indicated that the repression through Gal80p alone (in the absence of *MIG1*) is sufficient to cause a binary response. The binary response of the wild-type cells to glucose repression changes to graded when the cells are exposed to raffinose but not glucose, prior to subjecting the cells glucose repression.

It is not clear how and why cells have evolved different types of responses for repression and induction. Not only this, the growth conditions seem to alter the manifestation of a given type of response. While the exact mechanisms of this are

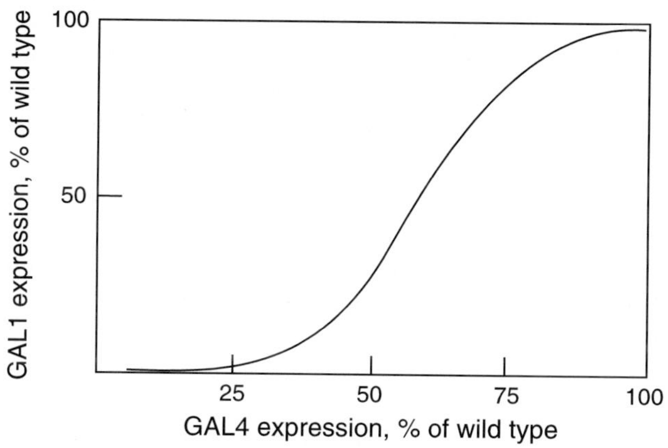

Fig. 7.2.3 *GAL1* promoter activity as a function of relative Gal4p concentration (adapted with permission from Griggs and Johnston 1991)

not clearly understood, many possibilities exist. For example, feedback loops, the multistep nature of signaling cascade, phosphorylation-dephosphorylation in signaling cascades have been shown to be responsible for nonlinear responses. There is evidence to suggest that just the genetic network alone can translate the graded input into a binary response. Further detailed analysis is required to unearth the mechanistic and physiological relevance of these observations.

7.2.5 GAL4 Expression is Repressed by Glucose

The presence of multiple pathways for glucose repression of *GAL* genes makes it difficult to evaluate the contribution of individual components to overall glucose repression. Since Gal4p is absolutely required for *GAL* gene induction, its repression alone is expected to play a significant role in glucose repression. It was evaluated by expressing different levels of Gal4p in strains deleted for *GAL4* and *GAL80* to ensure that extent of *GAL1* expression is a consequence of a change in Gal4p protein concentration. For this purpose, a series of strains were constructed bearing mutations in *GAL4* promoter that decreased the constitutive expression of *GAL4* to different extents as compared to the wild-type strain. The expression of *GAL4* in these mutant strains was indirectly determined by assaying the reporter chloramphenicol acetyl transferase by the wild-type (100%) and *GAL4* mutant promoters (expressed as percentage of wild-type).

The status of *GAL* gene expression in the strains expressing different levels of Gal4p was determined by monitoring the β-galactosidase driven by the *GAL1* promoter. Expression of β-galactosidase from *GAL1* promoter in a strain bearing wild-type *GAL4* but lacking *gal80* was considered 100%. It is clear from the pattern that a five-fold decrease in GAL4 (from 100 to 15%) resulted in a decrease in β-galactosidase expression by at least 40-fold (from 100 to 2.5%, Fig. 7.2.2). Note that the amount of GAL4 expressed is relative, and not in absolute terms (that is the absolute concentration of Gal4p in these strains was not determined). Nevertheless, these results tell us that the *GAL1* expression is quite sensitive to change in Gal4p concentration. β-galactosidase expression decreased as a function of the decrease in Gal4p protein in a sigmoid fashion. This observation is consistent with the previous observation that Gal4p binds UASg in a cooperative manner. That is, a small decrease in Gal4p tilts the balance from ON, to a state of OFF. This analysis was the starting point for developing a quantitative model of *GAL* gene expression (see section 8.2.2).

References

Adams BG (1972) Induction of galactokinase in *Saccharomyces cerevisiae*: kinetics of induction and glucose effects. J Bacteriol 111:308–315

Biggar SR, Crabtree GR (2001) Cell signaling can direct either binary or graded transcriptional responses. EMBO J 20:3167–3176

Carlson M (1999) Glucose repression in yeast. Curr Opin Microbiol 2:202–207

Ferrel Jr, JE (1996) Triping the switch fantastic: how a protein kinase cascade can convert graded inputs into switch-like outputs. Trends Biochem Sci 21:460–466

Gancedo JM (1992) Carbon catabolite repression in yeast. Eur J Biochem 206:297–313

Griggs DW, Johnston M (1991) Regulated expression of the GAL4 activator gene in yeast provides a sensitive switch for glucose repression. Proc Nat Acad Sci 88:8597–8601

Johnston M (1999) Feasting, fasting, and fermenting. Trends Genet 15:29–33

Klein CJ, Olsson L, Nielsen J (1998) Glucose control in *Saccharomyces cerevisiae*: the role of MIG1 in metabolic functions. Microbiology 144:13–24

Nehlin UO, Carlberg M, Ronne H (1991) Control of yeast *GAL* genes by MIG1 repressor: a transcriptional cascade in the glucose response. The EMBO J 10:3373–3377

Novick A, Weiner M (1957) Enzyme induction as an all or none phenomenon. Proc Natl Acad Sci USA 43:553–566.

Ronne H (1995) Glucose repression in fungi. Trends Genet 11:12–17

Rossi FM, Kringstein AM, Spicher A, Guicherit OM, Blau HM (2000) Transcriptional control: rheostat converted to ON/OFF switch. Mol Cell 6:723–728

7.3 Fine Regulation of *GAL* Genetic Switch

7.3.1 Introduction

Under natural conditions, concentrations of nutrients keep fluctuating in a dynamic fashion. Free-living organisms have developed intricate genetic regulatory mechanisms to cope with this constantly changing environment, so much so that even the genes under the same genetic regulatory unit may have to be differentially regulated in time, space, and magnitude. For example, although *GAL3* and *GAL80* are members of the *GAL* family, their expression pattern is vastly different than the other members. Difference in the expression pattern exists even between the enzyme of the metabolic pathway. How does a cell achieve such complex tasks?

What features of molecular interaction dictate this fine regulation? Regulation of *GAL* gene transcription provides some insights.

7.3.2 Basal and Induced Expression

There are two notable differences in the magnitude of expression between members of the *GAL* gene family. *MEL1, GAL3*, and *GAL80* contain one UASg in their promoter, and are known to have a Gal4p-dependent basal expression but the magnitude of induction response of these genes is high to moderate. The rest of the members of the *GAL* family such as *GAL* cluster have two UASg, with undetectable basal expression but the induction level is very high. Attempts to explain these differences based on the differences between sequences of UASg elements or the in vitro binding parameters have not yielded satisfactory results. To address this issue, strains were engineered to express ß-galactosidase from promoter containing either one or two UASg. The UASg elements used in the above study were identical to the UASg elements present in the promoter of *MEL1*. Basal ß-galactosidase expression was significantly reduced and at the same time the induced expression was significantly higher when two UASg elements are present in the promoter as compared to one (Table 7.3.1). This experiment demonstrates that the number of UASg elements has a profound effect not only on basal but on induced expression as well.

Under conditions of limiting amount of Gal4p, increasing the binding elements from one to two increase the probability of Gal4p interaction to give rise to higher induction response. What is the mechanistic basis for increasing the occupancy of the promoter consisting of two UASg under conditions of limited Gal4p in the cell? As mentioned before, Gal4p binds the UASg elements in a cooperative manner (see section. 8.2). That is, if a Gal4p dimer occupies the first UASg, the probability of the second dimer occupying the neighboring UASg is increased by at least an order

Table 7.3.1 β-galactosidase expression from promoters bearing different numbers of UASg (data obtained with permission from Melcher and Xu 2001)

Promoter elements	Basal expression	Induced expression	Basal (%)
One site of *MEL1*	84	570	12
Two sites of *MEL1*	14	2,300	0.2
Four sites of *GAL1*	43	5,000	0.7
Vector control	10	10	

Table 7.3.2 GAL4-dependent basal expression of α-galactosidase expression is due to incomplete repression by GAL80 (data obtained with permission from Melcher and Xu 2001)

Genotype	Uninduced	Induced
GAL4GAL80	16	880
gal4GAL80	<1	<1
GAL4 +multicopy *GAL80*	1	180

of magnitude. Therefore, cooperativity plays a crucial role in higher induction level observed for genes with two UASg as compared to one UASg.

Basal expression from a promoter containing one binding site is expected to be less compared to the promoter containing two binding sites, which is not what is observed. How come *GAL* genes with promoters containing two UASg have low basal despite having higher probability of interacting with Gal4p? A possible reason for this comes from the observation that over-expression of Gal80p from multicopy plasmid reduced (but did not completely abolish) the induced expression of *MEL1* gene containing single UASg, indicating that the concentration of Gal80p present in a wild-type strain is limiting. It has been demonstrated that Gal80p dimerizes in vivo. It has also been proposed that a weak dimer–dimer interaction exists that is stabilized by adjacent Gal4p dimmers. This Gal80p dimer–dimer interaction is responsible for the low basal expression of promoters with two binding sites. Such an interaction is not possible in genes with one binding site. Thus, by precisely titrating the amount of Gal4p, Gal80p and the binding elements, yeast has optimized the basal and induced expression.

Temporal regulation of GAL cluster: Another noteworthy feature is the difference in the time kinetics of induction between members of the *GAL* cluster. If at all, a temporal induction is necessary within the otherwise coordinately induced *GAL* cluster, galactokinase expression should have occurred first as it catalyses the first reaction in the pathway. However, it has been observed by independent groups that the transcription of *GAL7* precedes that of *GAL1* in response to galactose. What is the mechanistic basis of this temporal regulation? A likely possibility is that the *GAL1* promoter is less sensitive to the inducing signal compared to the *GAL7*. That is, as Gal80p is withdrawn from the nucleus during the early phase of induction, Gal80p present on the *GAL7* promoter appears to dissociate earlier than Gal80p present on *GAL1*

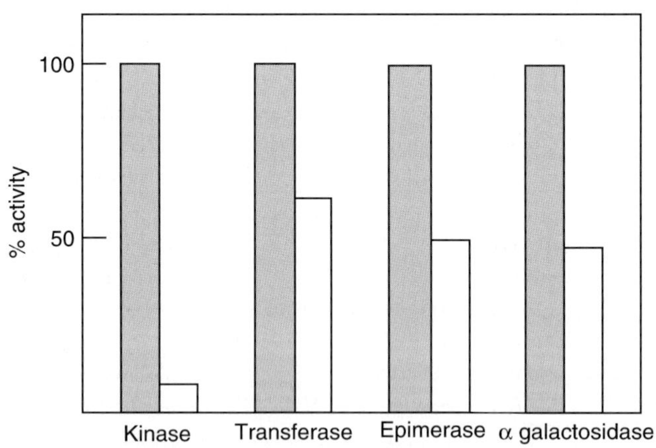

Fig. 7.3.1 Comparison of constitutive expression of *GAL* enzymes caused by over-expression of wild-type and Y57W mutant Gal3p. Constitutive enzyme activity of yeast strain transformed with multiple copies of wild-type GAL3 (*shaded bars*) or Y57W variant of *GAL3* (*open bars*). Constitutive induction of *GAL* enzymes by multiple copies of wild-type *GAL3* is taken as 100%

promoter. If this is true, then the steady-state expression of *GAL7* would be higher than *GAL1* or *GAL10* at low galactose concentrations. However, at saturating galactose concentration, no difference in expression between these genes could be observed. Since galactose is also consumed during induction, the difference in the steady-state expression that might exist between *GAL7 GAL1* and *GAL10* at low galactose concentration is difficult to discern as galactose is quickly consumed.

The difficulty addressed above was overcome by using a mutant derivative of Gal3p. We know that over-expression of Gal3p induces the expression of GAL genes to 40% of the fully induced levels. As mentioned before, over-expression of the Y57W mutant Gal3p conferred lower constitutive induction as compared to the wild-type. Constitutive expression of *GAL1, 7, 10*, and *MEL1* was separately measured in transformants over-expressing wild-type and the T57W mutant *GAL3* separately. Constitutive expression of *GAL7, 10*, and *MEL1* was approximately 50% while *GAL1* was less than 5% of what a wild-type Gal3p would induce. This suggested that for the given amount of Gal80p sequestered by T57W mutant, the response of *GAL 7, 10*, and *MEL1* is at least ten times higher than *GAL1*. That is, the *GAL* promoters have intrinsic differences in sensitivities for withdrawal of Gal80p from the nucleus.

7.3.3 Post-Translational Modification of GAL4 Protein

Gal4p is a phosphoprotein, and under non-inducing conditions, two distinct phosphorylated species can be detected by their differential mobility on SDS PAGE. Concomitant with transcriptional activation, Gal4p is further phosphorylated and the third lower migrating species can be identified. Of the many sites, phosphorylation of S699 is crucial for fine regulation. *SRB10* gene product, a RNA polymeraseII-holoenzyme associated cyclin-dependent protein kinase phosphorylates S699. The significance of this phosphorylation was not clear for some time. It turned out that a strain bearing a S699A allele instead of the wild-type *GAL4* cannot activate *GAL* when galactose concentration is 0.02%. Similarly, a strain bearing the S699A allele cannot activate *GAL* genes, even in the presence of 2% galactose if Gal3p function is absent. That is, the *gal3srb10* strain is non-inducible regardless of galactose concentration. Thus, phosphorylation of S699 is indispensable if galactose-dependent signal is weak. It has been observed that *SRB10* kinase is not synthesized under conditions of nutrient limitation. By incorporating the phosphorylation of S699 residue through *SRB10* gene product, additional constraints are imposed to ensure that the *GAL* system is not turned on when the nutritional status is not congenial.

More recently, Gal4p activity is also shown to be regulated by the ubiquitin degradation system through F-box proteins Grr1p and Dsg1p. Under non-inducing conditions, the Gal4p protein level is maintained by the Grr1p. In contrast, Dsg1p acts as a positive regulator of transcription by degrading the transcriptionally active Gal4p. Dsg1p is recruited to the UASg only during activation. Deletion of Dsg1 abolishes *GAL4* function indicating the importance of turnover of Gal4p in transcriptional regulation. Phosphorylation of Gal4p appears to mark the protein for Dsg1 mediated proteolysis.

The above examples convey an important message that cells have evolved stringent regulation by precisely adjusting the concentration of regulatory proteins, their affinities and the number of binding elements. Even the temporal induction of the *GAL* cluster seems to be important in that the toxic galactose-1-phosphate, the product of galactokinase catalyzed reaction, does not accumulate even transiently during galactose metabolism.

References

Ding WV, Johnston A (1997) The DNA-binding and activation domain of Gal4p are sufficient for conveying its regulatory signals. Mol Cell Biol 17:2538–2459

Lakshminarasimhan A, Bhat PJ (2005) Replacement of a conserved tyrosine by tryptophan in Gal3p of *Saccharomyces cerevisiae* reduces constitutive activity: implication for signal transduction in the *GAL* regulon. Mol Gen Genom 274:384–393

Novick A, Weiner M (1957) Enzyme induction as an all or none phenomenon Proc. Natl. Acad. Sci. USA 43:553–566

Melcher K, Xu HE (2001) Gal80-Gal80 interaction on adjacent Gal4p-binding sites is required for complete *GAL* gene repression. EMBO J 20:841–851

Muratani M, Kung C, Shokat KM, Tansey WP (2005) The F-box protein DSG1/MDM30 is a transcriptional co-activator that stimulates Gal4 turnover and co-transcriptional mRNA processing. Cell 120:887–899

Mylin LM, Bhat PJ, Hopper JE (1989) Regulated phosphorylation and dephosphorylation of GAL4, a transcriptional activator. Genes Dev 3:1157–1165

Gregar IH, Proudfoot NJ (1998) Poly(A) signals both transcriptional termination and initiation between the tandem GAL7 and GAL10 genes of *Saccharomyces cerevisiae*. EMBO J 17:4771–4779

Rohde JR, Trinh J, Sadowski I (2000) Multiple signals regulate transcription in yeast. Mol Cell Biol 20:3880–3886

Sadowski I, Costa C, Dhanawansa R (1996) Phosphorylation of Gal4p at a single C-terminal residue is necessary for galactose inducible transcription. Mol Cell Biol 16:4879–4887

Sadowski I, Niedbala D, Wood K, Ptashne M (1991) GAL4 is phosphorylated as a consequence of transcriptional activation. Proc Nat Acad Sci USA 88:10510–10514

St. John TP, Davis RW (1981) The organization and transcription of galactose gene cluster of *Saccharomyces*. J Mol Biol 152:285–315

Traven A, Jelicic B, Sopta M (2006) Yeast Gal4: a transcriptional paradigm revisited. EMBO Rep 7:496–499

Vashee S, Xu H, Johnston SA, Kodadek T (1993) How do Zn2 Cys6 proteins distinguish between similar upstream activation sites? J Biol Chem 268:24699–24706

Verma M, Bhat PJ, Venkatesh KV (2005) Steady state analysis of glucose repression reveals hirarchical expression of protein under Mig1p control in *Saccharomyces cerevisiae*. Biochem J 288:843–849

Xu HE, Kodadek T, Johnston, SA (1995) A single GAL4 dimer can maximally activate transcription under physiological conditions Proc Nat Acad Sci USA 92:7677–7680

Chapter 8
Paradigmatic Role of Galactose Switch

8.1 *GAL* Regulon and Genomics

8.1.1 Introduction

In classical genetics, inferences are based on the phenotypic differences between a wild-type and a mutant organism. This old-fashioned systems biology continues to provide deep insights into the genetic basis of fundamental biological processes mainly at the phenomenological level. This limitation was partly overcome by the revolutionary developments in recombinant DNA technology (Chap. 5). With this, it was possible to isolate, manipulate in vitro, and reintroduce a gene into the organism to determine and evaluate its effects on the performance at the phenotypic level. The impact of these combined approaches in deciphering the mechanistic details of molecular interactions has been impressive. However, biological problems are too challenging to be understood, even with these powerful approaches. Nevertheless, the classical and molecular genetic approaches culminated in a large body of data including the whole genome sequence of different organisms resulting in a paradigm shift in experimental biology beyond expectations.

In no other experimental organism genomic approach has been so successfully used than in *S. cerevisiae*. This is because yeast has been the vanguard of biochemical, genetic, and cell biological studies. An existing wealth of knowledge of yeast biology was put to great advantage in the context of sequence information. One of the immediate outcomes of the genome sequence is "expression profiling" using micro-array analysis. This technique has become quite routine and has yielded valuable information not only in yeast but also in almost all organisms whose genome sequence is known.

With the availability of complete genome sequence of yeast, strains bearing deletions in every known gene were generated. Here, each gene was deleted from start to stop codon by substituting with a deletion cassette consisting of gene of kanamycin flanked on either side by unique 20-nucleotide sequences that act as bar codes (see Box 8.1.1). A total of 5,916 strains were generated, each one bearing a mutation in a single gene. This constitutes approximately 96.5% of the total genes.

P.J. Bhat, *Galactose Regulon of Yeast*.

Box 8.1.1 *Disruption strains*

A battery of yeast strains each bearing disruptions of single ORF were constructed using deletion cassette specific for each ORF. The deletion cassette has a dominant selection marker for Geneticin (KAN). Using the appropriate upstream and downstream primers it is possible to identify and quantify the fraction of cells in a population by amplifying the unique tags associated with each ORF.

The 18mer present on the 5~ end of the 74mer primers is homologous to the specific ORF. The 18mer present on the 3~ end is homologous to the KAN gene, which directs the first PCR reaction. In the second PCR, the homology region for homologous recombination is increased by using a 45mer homologous to the 5~ and 3~ end of the ORF.

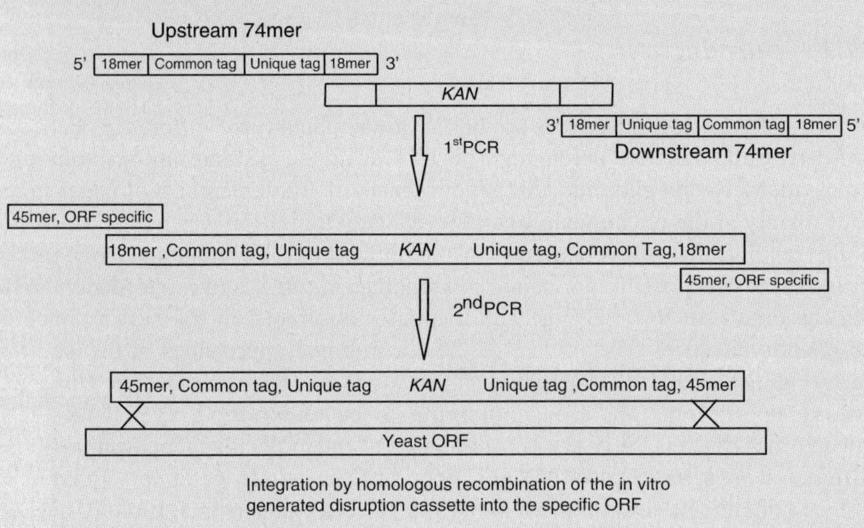

Fig. 8.1.4 Schematic representation of strategy for the systematic deletion of ORF strategy

Further analysis of these deletion mutants revealed that 18.7% of the genes (1,105 genes) are essential for viability. Of the viable deletion strains, 15% exhibit slow growth phenotype in rich medium at 30 °C. The genes representing this 15% include ribosomal proteins, mitochondrial proteins, and proteins involved in respiration. A significant number of ORFs in yeast still remain to be characterized and no single method can be prescribed to functionally characterize these genes. Here I shall discuss the genomic approach of functional analysis using *GAL* system as a paradigm, which gives us a glimpse of its ramifications in understanding the biology at a global level.

8.1.2 *Functional Profiling of Fitness*

Adaptive responses to changes in extracellular environment have been studied in great detail in yeast. Growth fitness under changing environmental condition is an emergent property that reflects the sum total of many biological processes. Traditional technology of the "gene-by-gene" approach is not suitable for identifying the contribution of gene(s) for growth fitness. Functional profiling was undertaken to identify the contribution of genes required for fitness of yeast growing on galactose. The specific sequence tag associated with each deletion locus allows the experiment to be analyzed in parallel. A pool of strains, each carrying a deletion in different genes, is allowed to grow on galactose over a period of time. Reduction in the number of cells defective for a given gene as a function of time points out that the gene in question is necessary for growth fitness in galactose. A decrease in the number of cells bearing deletions was quantitated by amplifying the unique bar code characteristic of each deletion strain followed by hybridization to an oligonucleotide array of the complementary bar code sequences. The reasoning is that the importance of the gene for fitness under the experimental condition is inversely related to the representation of the strain deleted for a particular gene.

The above approach identified *MSN2, FET3, YDR290W, ATX*1, *YNL077W, YDR-269C, GEF1, YML090w, YKL037w* in addition to the previously known genes, which upon deletion reduced fitness in galactose medium. In subsequent experiments, strains bearing the above deletions grown individually in galactose medium showed a growth defect of 41–91% of wild-type growth (Fig. 8.1.1a).

The fitness-profiling data can be superimposed on the expression-profiling data as both interrogate the whole genome and are expected to show the following correlation. In general, a gene that is highly expressed in galactose is expected to show a growth defect in galactose if it is deleted. For example, *GAL1* is highly expressed in galactose and if it is deleted, the strain will have a growth defect in galactose. The expression profile of 4,682 genes was monitored in the presence and absence of galactose. Of these, 98 and 84 genes, were significantly up- and down-regulated, respectively. What was surprising was that many of the genes that exhibited fitness defect upon deletion did not exhibit a significant increase in expression. For example, less than 7% of the genes that showed fitness defect upon deletion exhibited a significant increase in mRNA (Fig. 8.1.1b) expression in galactose. Expression of some genes required for fitness did not increase in expression while in other cases, expression increased but deletion did not contribute for fitness (Fig. 8.1.1b). This study revealed that additional genes than those previously identified are required for fitness during growth.

The observation that a gene not highly expressed but is required for fitness is not surprising and can be rationalized on the basis that it exerts its role because of post-translational modification. However, the unexpected finding was that genes that are not required for fitness showed increased expression. This is a classic example of how a global approach can identify novel genes or phenomena that are otherwise not amenable for routine analysis.

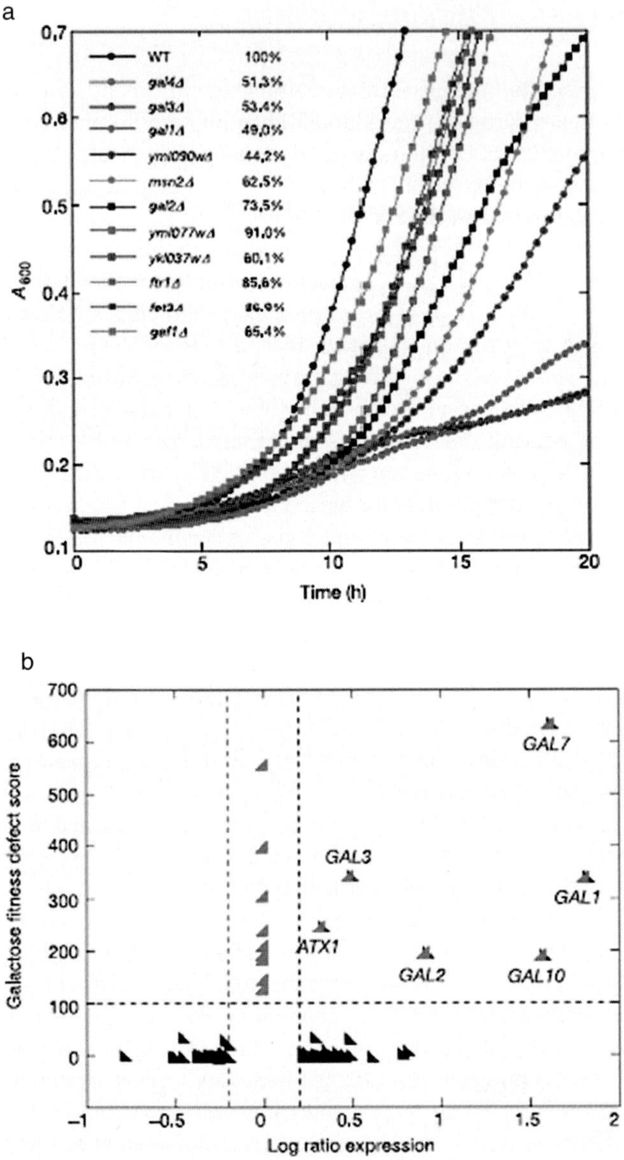

Fig. 8.1.1 Comparison of fitness defect and expression change. **a** Growth profile of individual strains identified to be defective in fitness in galactose as the sole carbon source. Percent of wild-type growth is represented in the *inset*. **b** expression change of *GAL1, 2, 3, 7, 10* and *ATX1* and fitness defect is correlated. Other genes that show significant fitness defect (above score 100) do not show a corresponding change in expression beyond –0.2 and 0.2 log expression ratio, respectively, (significant change is indicated by the two *dashed vertical lines*). Similarly, in those that show a significant change in expression ratio, the fitness defect score is zero (reproduced with permission from Gleaver et al. 2002)

8.1.3 Analysis of Genome-Wide DNA Binding

Reprogramming of gene expression occurs as cells move from one state to another. This is mainly due to regulation of transcription mediated by dedicated transcriptional activators such as Gal4p. Experiments conducted so far have revealed that only a few genes are activated in the presence of galactose, but genome sequence analysis revealed that approximately 300 Gal4p-binding sites were spread throughout the genome. In fact, one such site was present within the ORF of acetyl Co-A carboxylase. Elimination of this site does not lead to a defect in galactose growth, indicating that it has no functional role. What determinants prevent the unnecessary occupancy of such sites by Gal4p? As discussed before, interaction with other proteins might provide the additional specificity required.

To determine the functionally relevant Gal4p-binding sites in the whole genome, the genome was scanned by chromatin immunoprecipitation assay. For this purpose, C-terminal myc tagged *GAL4* was integrated into its native genomic locus by homologous recombination. The functional UASgs were immunoprecipitated, amplified, and labeled to interrogate the microarray of intergenic regions (Fig. 8.1.2). This approach identified *MTH1, PLC10* and *FUR4* as the additional genes whose promoters were recognized by

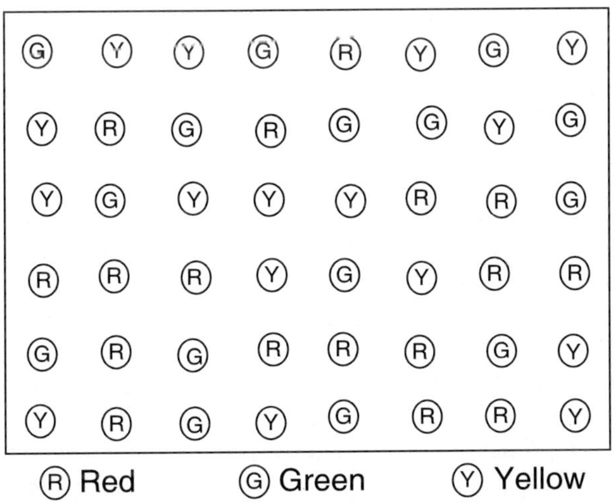

Fig. 8.1.2 Schematic illustration of a of section of an image of microarray data. Each spot corresponds to a unique intergenic region. The slide containing 6,361 interegenic region was interrogated by a mixture of two sets of PCR-amplified labeled probes obtained from fragments enriched by immunoprecipitation (in this example, Gal4p bound DNA was precipitated) using antibodies (labeled with cyanin 5, red) or its corresponding control (labeled with cyanis 3, green). A given spot would appear as either red or green or different shades in between these two extremes. The intensities of these colors are a reflection of the relative concentration of these two labeled probes in the population. A *red spot* represents higher abundance of cyanin 5 labeled probe while a *green spot* represents higher abundance of cyanin 3 and a *yellow spot* indicates that both are in equal concentration. For example, a spot corresponding to GAL1-10 intergenic region shows red, clearly indicating that the mixture has a higher concentration of cyanin 5 labeled GAL1-10 probe

Gal4p through UASg. In independent experiments, expression of these genes was shown to be induced by galactose in a Gal4p-dependent manner.

FUR4 encodes uracil permease, whose function might be to provide intracellular pool of uridine, which is required for the synthesis of UDPglu, the precursor for galactose metabolism. *MTH1* is a repressor of *HXT* genes which code for hexose transporter. This indicates that galactose specifically represses hexose transporter not required for growth on galactose. *PLC10* codes for cyclin, which appears to repress glycogen synthesis. Induction of *PLC10* by galactose could maximize the flux of galactose for energy production than diverting it to glycogen.

8.1.4 Genomic Approach for Network Analysis

The purpose of this exercise was to establish the proof of principle that genome-wide expression analysis of both mRNA and protein levels can be used to unearth the network of interaction. Since a detailed understanding of *GAL* gene regulation was already available, the results of perturbation of this network can easily be predicted. The idea then was to see whether similar predictions can be arrived at by observing the change in expression pattern by subjecting the system to perturbations. As many as 20 genetic and environmental perturbations (wild-type, *gal1, gal2, gal3, gal4, gal5, gal6, gal7, gal10* and *gal80* in presence and absence of galactose constitute a total of 20 different experimental perturbations) were introduced and genome-wide expression of mRNA and protein was determined. This data was then analyzed in the context of the known model of *GAL* genetic switch and genome-wide protein–protein interaction data already available.

Of the total 6,200 yeast genes, transcription of 997 genes showed consistent differences in expression levels from reference (wild-type grown in galactose medium) under one or more perturbations. The 997 genes were then separated into 16 groups, where each group contained genes with similar expression profile over all perturbations. Of the 16 clusters, clusters 1, 2, and 3 together contained all seven genes with established *GAL4*p-binding sites. Of the remaining 87 genes in these clusters, nine had putative Gal4p-binding sites not previously identified. These nine genes represented functions involved in glycogen accumulation, protein metabolism, and unknown functions. For example, *PLC10* was identified independently during a genome-wide search for the genes whose regulation is dependent on Gal4p (see previous section).

This study was also extended to determine whether the changes observed in mRNA levels also reflect a corresponding change in protein levels. For this purpose, the protein abundance between wild-type +galactose and wild-type −galactose was measured using the isotope-coded affinity tag method in tandem with mass spectrometry. Using this technique, only 229 proteins abundance could be detected. This included all the *GAL* proteins that showed a difference in the expression. Of the 30 proteins that showed clear changes, mRNA levels for 15 did not show any corresponding changes. The correlation between the changes in the levels of protein and mRNA in wild-type cells with and without galactose was in the range of 0.48. In some cases, the increase in proteins was observed without a concomitant increase in mRNA and vice versa. The slope of the log-log plot of the protein expression ratio against the mRNA expression ratio is of the order of 0.301, which means that a ten-fold change in mRNA results in a two-fold increase in proteins (Fig. 8.1.3).

A result that was not predicted by the known regulatory model was obtained when the system was perturbed by introducing mutations at *GAL7* and *GAL10* loci (two of the 20 perturbations). These two perturbations decreased the expression of other *GAL* genes when galactose was present. It was hypothesized to be due to the accumulation of galatcose1-phosphate due to a metabolic block imposed by *GAL7* or *GAL10* defect. This hypothesis was found to be true, since disruption of *GAL1* in combination with *GAL10* did not result in the changes in the expression of *GAL* gene seen only when strains were disrupted for *GAL7* or *GAL10* were grown in

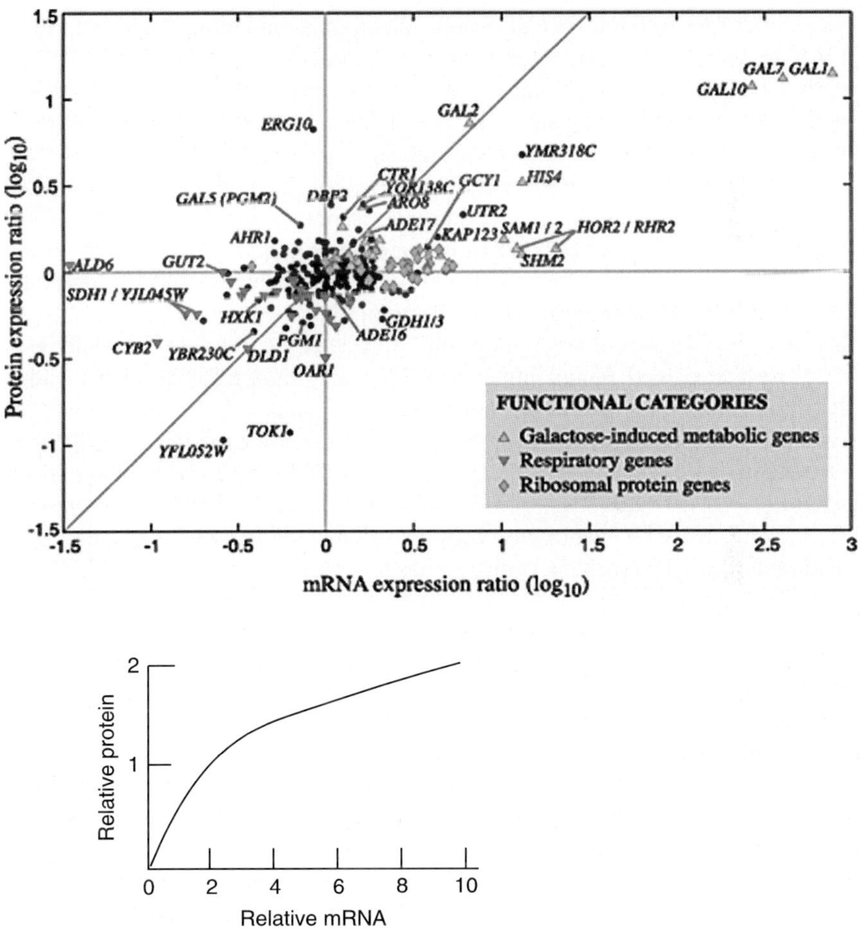

Fig. 8.1.3 Scatterplot of protein expression versus mRNA expression. Ratios of wild-type +galactose to wild-type −galactose protein expression measured for each of 289 genes are plotted against the corresponding mRNA expression ratios (reproduced with permission from Idekar et al. 2001). In the lower panel, the relative levels of proteins are plotted against relative levels of mRNA on linear co-ordinates (adapted with permission from Fell 2001)

Box 8.1.2 *Nucleic acid hybridization assays*

The discovery of nucleic acid hybridization, between DNA:RNA, first discovered in 1961 by Hall and Spiegelman, is the basis for many of the techniques that use this principle. Southern and Northern hybridization techniques take advantage of the cloned piece of DNA as a hybridization probe to screen for the presence of homologous sequences within a complex mixture of DNA or RNA. This technique was later adopted for screening plasmid or phage clones containing cloned pieces of DNA by colony hybridization. With this technology, the complex nucleic acid mixture is generally separated through electrophoresis, or by plating for individual colonies or plaques. The nucleic acid is then transferred to a nitrocellulose filter which is then probed by the cloned piece of DNA. Differential plaque hybridization technique used to clone the *GAL* cluster is a variation of the above technique.

In micro-array, unseparated heterogeneous population of nucleic acid obtained from two sets of samples to be compared are differentially fluorescent labeled. One sample is labeled with cyanin 5 and the other with cyanin 3. These two sets of labeled molecules are allowed to hybridize simultaneously to a known set of ORFs, or oligonucleotides corresponding to ORFs, or intergenic regions or unique tags, (bar codes) that are deposited on a glass slide. This technology allows miniaturization in that approximately 10,000 different sequences can be spotted in an area of 2×2 cm. The advantage of this technique is that a large number of target molecules present in the sample can simultaneously be interrogated. For example, cDNA prepared from mRNA isolated from two populations are separately labeled using cyanin 5 (red) and cyanin 3 (green), mixed and allowed to hybridize a microarray containing the DNA. Depending on the relative abundance of an mRNA, the corresponding spot would be red (if the mRNA is more in sample labeled with cynin 5), green (mRNA labeled with cyanin 3 is more) or yellow (equal in both the samples). The signal per spot (each spot corresponds to a gene or intergenic region, etc.) gives an indication of the relative amount of the probes present in the two populations of the samples. This quantitation can be simultaneously done for as many DNA spots present in the microarray chip.

galactose. While it was known that galactose-1-phosphate accumulation causes toxicity, its effect on gene expression was not revealed by the previous studies.

One other unanticipated observation was the slow growth of the *gal80* mutant in medium lacking galactose. It was hypothesized that the slow-growth in the absence of galactose is due to the constitutive expression of *GAL* genes. This idea was tested by comparing the growth of *gal4gal80* strain in the absence of galactose. The reasoning is that the constitutive expression of *GAL* genes in a *gal80* strain would be abolished if *GAL4* is also deleted. As expected, the doubling time and the expression profile of *gal4gal80* double deletion strain were more similar to the *gal4* deletion strain than *GAL80* deletion strain. This indicated that the constitutive expression of *GAL* genes in *gal80* strain is the cause of the observed slow growth of the gal80 strain in the absence of galactose.

Here, we have discussed three distinct genomic approaches wherein prior knowledge of galactose metabolic regulation was combined with information available at the gnomic scale. Genomic approaches tell us that it is possible to study a system for which limited prior information is available. Genomic approaches can unearth the network of gene interactions responsible for producing the phenotype, which is otherwise not amenable for routine analysis.

References

Cornish-Bowden A, Cardenas M (2001) Complex network of interactions connect genes to phenotypes. Trends Biochem Sci 26:463–464

Fell D (2001) Beyond genomics. Trends Genet 17:680–682

Glaever et al (2002) Functional profiling of *Saccharomyces cerevisiae* genome. Nature 418:387–391

Hall BD, Spiegelman S (1961) Sequence complimentarity of T2 DNA and T2 specific RNA. Proc Nat Acad Sci USA 47:137–146

Hoheisel JD (2006) Microarray technology: beyond transcript profiling and genotype analysis. Nature Reviews 7:200–209

Idekar T et al (2001) Integrated genomic and proteomic analyses of a systematically perturbed metabolic network. Science 292:929–933

Ren B et al (2000) Genome-wide location and function of DNA-binding proteins. Science 290:2306–2309

8.2 *GAL* Regulon and Systems Biology

8.2.1 Introduction

Biological systems are highly organized dynamic molecular ensembles that respond in a pre-determined manner to fluctuations in intra- and/or extra-cellular conditions. Molecular interactions governed by structural complementarity are at the heart of such varied biological manifestations starting from catalysis of diverse chemical reactions, cell–cell communications to complex phenomenon such as memory. A deterministic outcome as complex as life can emerge only if these interactions are precisely coordinated in time, space, and magnitude. This coordination is instituted by the network of interaction which determines, for example, why an ant does not grow to the size of an elephant. This clearly means that biological processes are self-regulatory in nature. Obviously, besides structural complementarity, other mechanisms such as feedback regulation and cooperativity play a significant role in maintaining order in the midst of apparent disorder.

Historically, feedback regulation was identified in biosynthetic pathways. Here, the activity of the rate-limiting enzyme is controlled by the end product of the biosynthetic pathway. This is a feedback loop that imposes a “self-restraint” that controls the synthesis of the end product within a concentration range that is just sufficient for

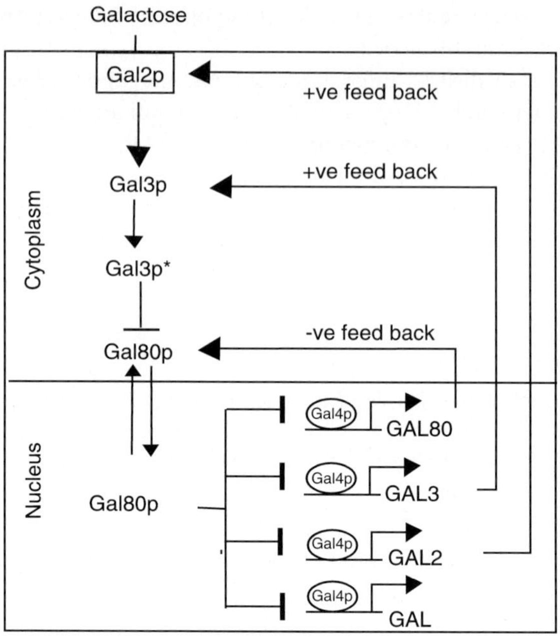

Fig. 8.2.1 Schematic representation of *GAL* genetic network. *Arrows* point out activation while the *thick vertical line* indicates inhibition. Transformation of Gal3p to its active form is indicated by the *

the cell growth. Cooperativity was initially observed in hemoglobin and later in many of the metabolic enzymes. The significance of cooperativity in biological systems can be appreciated from the Biblical parable of the rich and the poor: "For unto everyone that hath shall be given, and he shall have abundance; but from him that hath not shall be taken away even that which he hath". Genetic networks, too, employ feedback loops and cooperativity to orient themselves for defining the system property within a specific physicochemical boundary. I had previously alluded to feedback regulation (Fig. 8.2.1) and cooperativity at the phenomenological level. Here I shall discuss their biological significance in a little more detail.

8.2.2 Quantitative Basis of GAL Genetic Switch

Maximum *GAL* gene induction occurs in cells lacking *GAL80* gene. No matter how much galactose is present in the medium, a wild-type strain can only be induced to 80% of what a *gal80* mutant can. Genes with one UASg, such as *MEL1*, have a basal expression that cannot be completely repressed, even if Gal80p is over-expressed from a multicopy plasmid. We do not yet understand the quantitative basis of these regulatory features of the genetic switch that brings about such a stringent response. Here we shall discuss the working of *GAL* genetic switch from a quantitative perspective, which addresses some of the paradoxes discussed above.

If the performance of the switch has to be understood in its entirety, it is necessary to evaluate the concentrations of the components of the switch at rest and during induction. Here, the effort is to determine the concentration of Gal4p, Gal80p, and Gal3p in a wild-type cell and simulate the steady-state performance of the switch. This would give a glimpse of how the experimental data is related to a quantitative framework of regulatory interactions. Once a quantitative framework of the working of the switch is available, questions that are otherwise difficult to tackle using experimental regimen can be evaluated at the theoretical level.

It is possible to theoretically evaluate the extent of *GAL* gene expression for a given concentration of Gal4p. A fundamental assumption is that *GAL* gene expression at the population reflects the occupancy of Gal4p onto the DNA. That is, a 50% expression of *GAL* genes reflects an occupancy equivalent to 50%. The volume of a yeast cell is 70 μm^3. A total of seven promoters consisting of two UASg and three promoters with single UASg are considered. This translates to a concentration of 1.5×10^{-10} M for two Gal4p-binding sites and 6.0×10^{-11} M for single binding sites, respectively. A promoter with two binding sites is considered as one molecule with two binding sites for Gal4p. Since the binding constant of Gal4p to its cognate binding site has been experimentally determined to be 5×10^{-10} M, it is possible to determine the fraction of UASg that is occupied as a function of Gal4p dimer concentration. Wild-type Gal4p concentration is varied over a wide range, but satisfying the condition that six-fold change in Gal4p should result in a 40-fold change in expression. This condition is based on the observation discussed in Chap. 7.2.

For example, a decrease by five-fold from a starting Gal4p concentration of 1×10^{-5} M to 0.2×10^{-5} M would result in 100% occupancy of DNA. Similarly, an increase in Gal4p concentration from a starting low 1×10^{-15} M to 5×10^{-15} the occupancy will still be zero. These results indicate that the wild-type Gal4p concentration lies between 1×10^{-5} M and 1×10^{-15} M. By trial-and-error method, a wild-type Gal4p concentration of 5×10^{-9} M (Fig. 8.2.2) was arrived at, which satisfies the experimental result. That is, a five-fold decrease from 5×10^{-9} M, occupancy of Gal4p is decreased by 40-fold. This concentration translates to 295 molecules of Gal4p per cell. This result is reasonably close to the experimentally determined value, which is 200 molecules of Gal4p per cell.

Experimentally, it has been demonstrated that Gal4p binds UASg cooperatively. In the above simulations, a factor for cooperative binding to promoters with two binding sites was incorporated. For example, at low concentrations, Gal4p would not discriminate between UASg of *GAL1* and *MEL1* promoter, but as the concentration keeps increasing, Gal4p would bind *GAL1* promoter with higher probability as compared to *MEL1*. This binding occurs with a 30-fold higher affinity. At a concentration of 200 dimer Gal4p, the promoters containing the single binding sites are occupied only 70% as compared to 95% in case of *GAL1, 7* or *10*, corresponding to an expression level of 70 and 95%, respectively, (Fig. 8.2.2). This behavior is similar to what is experimentally observed, that is, *MEL1* expression is moderately high but not as much as the *GAL* cluster. This gives us a quantitative picture of the consequences of cooperative interaction.

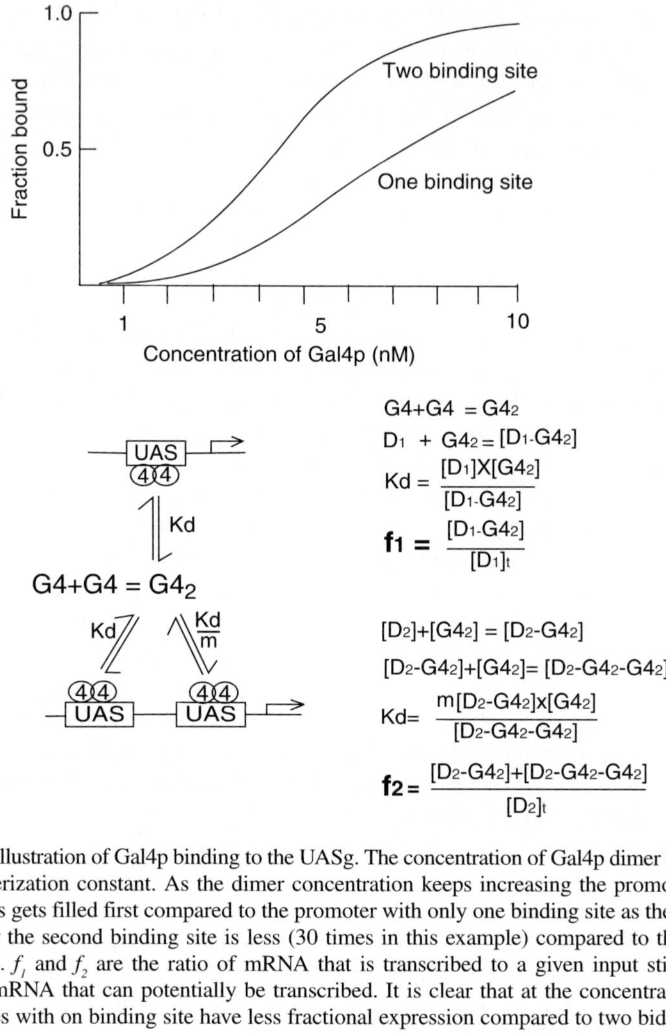

Fig. 8.2.2 Illustration of Gal4p binding to the UASg. The concentration of Gal4p dimer is determined by the dimerization constant. As the dimer concentration keeps increasing the promoter with two binding sites gets filled first compared to the promoter with only one binding site as the dissociation constant for the second binding site is less (30 times in this example) compared to the Kd of one binding site. f_1 and f_2 are the ratio of mRNA that is transcribed to a given input stimulus to the maximum mRNA that can potentially be transcribed. It is clear that at the concentration range of Gal4p, genes with on binding site have less fractional expression compared to two biding sites

Turning OFF and ON: Once the Gal4p concentration in a wild-type cell is determined, it is possible to estimate the concentration of Gal80p in a wild-type strain under non-inducing condition, given the affinity of Gal4p and Gal80p interactions is experimentally determined to be 5×10^{-9} M. Based on this, a concentration of 0.05 μM is sufficient to shut off all the Gal4p present in the cell. At this concentration of Gal80p, Gal4p present on single binding site is occupied by Gal80p only to the extent of 75%. This explains why genes with single UASg have a basal expression. Recall that under fully induced conditions, the Gal80p concentration increases five-fold over and above what is present before the induction.

The amount of Gal3p required to activate the system to 80% of the maximum in the presence of galactose was determined with an assumed dissociation constant of 5×10^{-10} M between Gal3 and Gal80p. This turns out to be 3.2 μM. This maximum

Gal3p occurs under fully induced conditions and not what is present before induction since Gal3p expression is also autoregulated. In calculating this concentration of Gal3p, the following parameters were considered. First, that Gal3p increases by 10-fold through autoregulation as the induction proceeds. Second, the Ks, that is the concentration of galactose required or half maximal induction, which was reported to be 0.1 M, based on experimental analysis. This parameter decides the extent of activation of Gal3p from inactive to active Gal3p, which keeps varying as the induction occurs.

A wild-type cell containing 0.005 μM, 0.05 μM and 0.25 μM of Gal4, Gal80p, and Gal3p, respectively, is poised to take off upon galactose addition (Fig. 8.2.3). From the initial Gal3p, Gal80p, and Gal4p concentrations, the extent of Gal4p not bound by Gal80p as a function of different fixed galactose concentration was generated at three different values of distribution coefficient of Gal80p. Recall that Gal80p is a nucleocytoplasmic shuttle protein and is sequestered in the cytoplasm due to its interaction with active Gal3p. As galactose concentration is increased, more and more Gal80p is sequestered in the cytoplasm. The distribution of free Gal80p is dependent on its distribution coefficient, which is not known.

Induction was simulated at Gal80p distribution coefficient of 10, 0.4 and 0.1. The simulated induction pattern agrees with the experimentally observed value only at a distribution coefficient of 0.4 (Fig. 8.2.3). Simulation results indicated that at saturating galactose concentration, 99.9% of 0.5 μM Gal80p is sequestered by 3.0 μM of Gal3p in the cytoplasm. Under these conditions, about 0.06% of 0.5 μM Gal80p protein is present in the nucleus. At this nuclear concentration of Gal80p, no more than 80% of Gal4p will be active. This is consistent with experimental observation. That is, in a wild-type strain no more that 80% of what a *gal80* strain can express can be achieved.

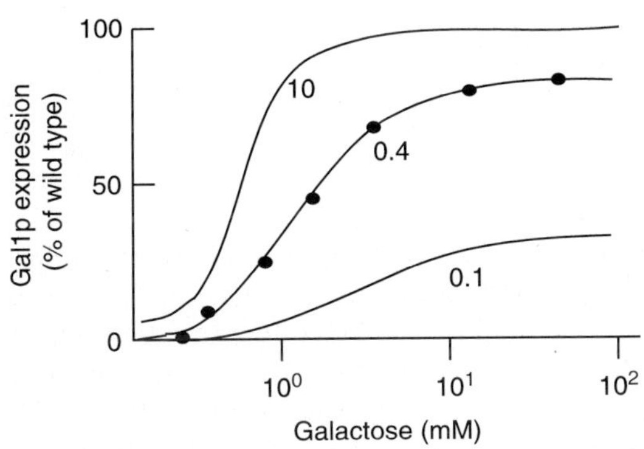

Fig. 8.2.3 Induction response of the *GAL* promoter as a function of galactose concentration. β-galactosidase (expressed form the *GAL1* promoter) expression was monitored at different fixed galactose concentrations (*filled circles*). A constant external galactose concentration was maintained using an automated input device. The response curves were generated by simulation, as discussed in the text at Gal80p distribution coefficient of 10, 0.4, and 0.1. The experimentally determined response curve is similar to the simulated one at Gal80p distribution coefficient of 0.4 (reproduced with permission from Verma et al. 2003)

What does simulation tell us? Based on the intracellular concentrations of Gal4p, Gal80p, and Gal3p, many experimentally observed results can be rationalized. For example, it has been demonstrated that the basal expression of *MEL1* cannot be completely suppressed by over-expression of Gal80p from a multicopy plasmid (see section 7.3.2). Simulation predicts that Gal80p should be over-expressed at least 1,000-fold more than what is present in a wild-type strain. Obviously, such a high concentration of Gal80p cannot be achieved under in vivo conditions. Secondly, in a wild-type cell, if the repression due to Gal80p has to be completely relieved to express maximally, then a concentration of 20 μM of Gal3p is required, which is not what is present in the cell. This explains why in a wild-type cell the induction reaches a limiting value of 80% of what is potentially possible. It is quite clear that the relative concentrations of these regulatory factors are an important determinant of the performance of the switch. With the tools of simulation in hand, it is possible to evaluate the role of negative and positive feedback loops of *GAL80* and *GAL3*. It turned out that disabling both of the loops did not alter the performance of the switch, however, disabling any one, and keeping the other intact, resulted in an abnormal performance of the switch. More recently, the significance of these loops has been studied in greater detail, which is discussed at the end of this chapter.

8.2.3 Long-Term Adaptation Revisited

Maintaining constancy in the relative concentrations of regulatory proteins as the cells divide ensures that the switch does not fortuitously turn ON when not required or vice versa. Cells have evolved regulatory mechanisms to ensure that such mishaps do not occur. To what extent can the switch withstand a fluctuation in the relative concentrations of these regulators without itself going out of control? In other words, how robust is the switch to changing concentrations of the regulatory components. For example, *GAL* switch should discriminate a drop in nuclear Gal80p concentration occurring only in response to galactose from a decrease due to normal fluctuations that occurs say during cell division. There has to be a trade-off between these two opposing demands if the switch has to fulfill its biological purpose.

Now that the concentrations of Gal4p, Gal80p, and Gal3p are known in the wild-type cell, it is possible to evaluate the consequence of a perturbation in the levels of any one of these regulators. Based on statistical considerations, random partitioning of Gal80p molecules would result in the following distribution. A total of 88% of the cells would have Gal80p within ±25% of the normal. That is, Gal80p concentration would vary between 0.0625 to 0.0375 μM in 88% of the cells. In the rest (6% of the cells), the concentration of Gal80p will be above this level and in the other 6% it will below this level. With this distribution, approximately four cells out of 1,000 would contain Gal80p less than 0.025 μM. What is the state of *GAL* gene induction in cells that receive less than 0.025 μM? Using the model, it was estimated that in cells expressing less than 0.025 μM of

Gal80p, *GAL1* is expressed at 0.05% of the fully inducible level, even if galactose is not present.

This analysis indicated that the long-term adaptation phenotype could as well be due to an inherent statistical fluctuation in the repressor concentration, which occurs during cell multiplication (Fig. 8.2.4). That 0.05% *GAL1* expression that occurs in a small fraction of cells is sufficient to initiate the induction cascade, provided galactose is present. In other words, only a fraction of cells in a population of *gal3* cells exposed to galactose eventually grow on galactose. In approximately four of 1,000 *gal3* cells the *GAL1* is expressed to the extent of 0.05%, due to a chance decrease of Gal80p below 0.025%. Because of this reason, if 1,000 wild-type and *gal3* cells are separately inoculated into a medium containing galactose, the wild-type cells would grow almost immediately, while the *gal3* cells would show a significant lag. This is because out of 1,000 *gal3* cells, only 4 are capable of growing on galactose and, as a consequence, the overall induction kinetics of cell multiplication will be slower as compared to the wild-type.

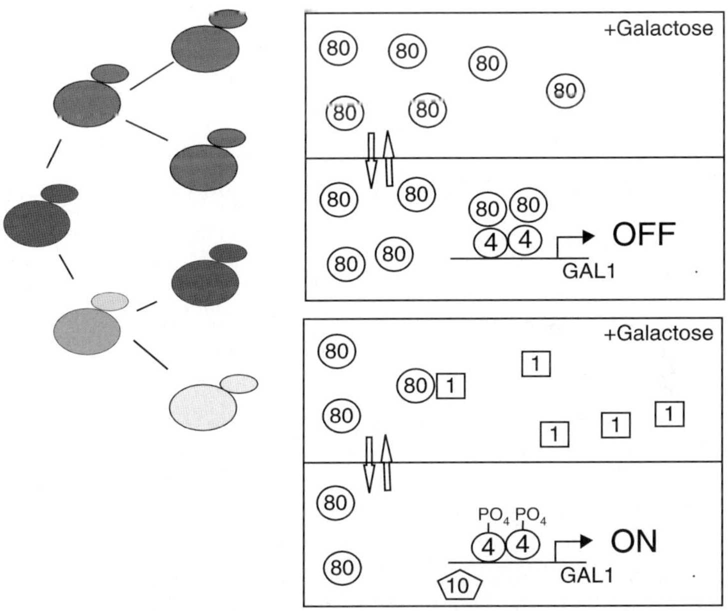

Fig. 8.2.4 Long-term adaptation phenotype is a consequence of unequal distribution of Gal80p. During cell growth on a non-inducing non-repressing carbon source, *gal3* cells would receive different amount of Gal80p. A small fraction of cells would receive Gal80p below the threshold required to keep the system in off state (*represented by light shade*) and in these cells *GAL1* would be expressed to the extent of 0.05% of the normal induced levels. Such cells turn on the system in the presence of galactose. This signaling pathway appears to be weak as a *SRB10* mutation in the background of *gal3* does not induce at all (see text for more details). Phosphorylation at serine 699 residue of Gal4p is essential for Gal1p expression induction pathway (reproduced with permission from Bhat and Venkatesh 2005)

8.2.4 Feedback Loops of GAL Regulon

Gal2p transports galactose from the medium into the cell. Unlike a wild-type strain, a *GAL2* mutant grows on galactose as the sole carbon source, provided its concentration is above 0.05%. In the absence of galactose, its induction is undetectable, but its expression is induced several hundred-fold in response to galactose, which means that *GAL2* is under the positive feedback loop (Fig. 8.2.1). The biological significance of this loop was evaluated experimentally by expressing *GAL2* independent of the positive feedback loop. Wild-type strains show a switch-like response to varying concentrations while, a *gal2* strain shows a linear response. If Gal2p is expressed constitutively from a promoter independent of the positive feedback loop, the linear response as observed in a *gal2* strain is retained, but shifts upwards compared to the linear response exhibited by *gal2* strain (Fig. 8.2.5). This implies that the positive feedback loop of Gal2p expression is an important element of the *GAL* genetic switch that is required for imparting a switch-like response. However, removal of the feedback loops of Gal3p and Gal80p did not result in the loss of the switch-like property.

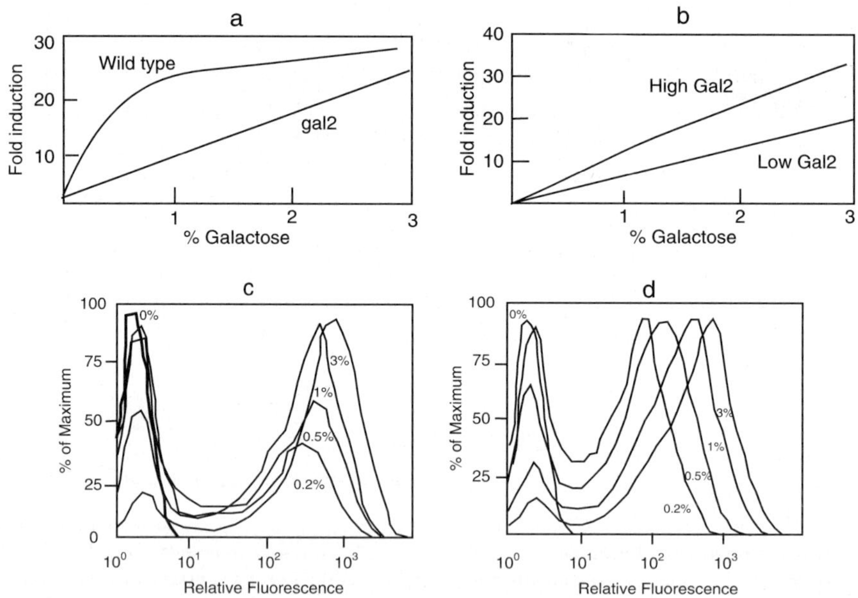

Fig. 8.2.5 Behavior of GAL2 feedback-disabled cell in response to galactose. **a** Fold induction of *GAL* gene expression as a function of galactose concentration in a wild-type and in a *gal2* disrupted strain. **b** Fold induction by altering Gal2p levels independent of the feedback loop. **c** Increase in binary response to galactose concentration in wild-type cells. **d** Increase in graded response in response to galactose in a *gal2* strain (adapted with permission from Hawkins and Smolke 2006)

Box 8.2.1 *Epigenetics*

According to the concept of classical genetics, what a cell or organism inherits from its parents is a repertoire of genes and not a state of gene expression. The clonal population of somatic cells (despite inheriting identical DNA sequences) end up being very different cell types. The new cell identity acquired during differentiation, without going through any changes in DNA sequence, are often stable and retain their differentiated features for many generations even under in vitro conditions of cell culture. Therefore, the fundamental questions are: How does functional diversity arise from genetically homogeneous or clonal population of cells? What is the origin of such a variability when two cells receive the same constellation of genetic determinants and how this state is being passed on from one cell to the other? Obviously, the answer has to do with mechanisms independent of the genes and yet influences the expression status of the genes.

Epigenetics is considered as a change in the state of expression of a gene that does not involve a mutation, nevertheless inherited in the absence of the signal or event that initiated the change. Epigenetic change does not introduce any change in the information content present in the genetic material. Long-term adaptation of *gal3* cells is an example of epigenetic phenomenon. A population of *gal3* cells is genetically identical to start with, but a small fraction of cells differ in their state with regard to their sensitivity to galactose. Such cells exposed to galactose start expressing *GAL* genes without the cells going through any genetic alterations. As discussed, a stochastic event resulting in a decrease in the Gal80p confers the increased sensitivity to galactose due to the expression of *GAL1* even in the absence of galactose. This low expression sustains itself through a positive-feedback loop and is maintained autonomously. A similar positive-feedback loop of *GAL3* has been shown to be responsible for the persistent memory again an example of epigenetic state. These mechanisms do not involve any special feature of the chromosome. Epigenetic mechanism allows the separation of the maintenance of the state of gene activity from the conditions for their initiation.

The response observed in the above experiments was at the population level and does not reflect the state of induction at the single-cell level. Single-cell analysis carried out in wild-type and *gal2* mutant indicated that the wild-type cells have a single steady state while *gal2* mutants have multiple steady states. That is, wild-type cells existed, either in fully induced or not induced at all. Increasing the galactose concentration only increased the fraction of fully induced cells (Fig. 8.2.5). In the *gal2* mutant, the population as a whole induced more and more as a function of galactose concentration. That is, the wild-type showed a binary response while the *gal2* mutant showed a graded response. At higher galactose concentrations, both strains exhibited similar population distribution.

In another independent study, feedback-disabled yeast cells were engineered wherein both *GAL3* and *GAL80* were expressed from constitutive promoters. That is, their expression was no more regulated by the feedback loop. Such cells exhibited a delayed initial response to galactose but attained higher steady-state level values within 5–6 h of induction as determined by the *GAL1* promoter-driven GFP expression. This data was consistent with the theoretical simulations. *GAL1*-driven GFP expression analysis indicated that approximately 80% of the wild-type cells responded within 2 h, while only 30% of the *GAL3 GAL80* feedback-disabled cells responded within the same time. Further analysis indicated that this difference is due to a cellular heterogeneity with respect to the Gal80p concentration. The population of cells having higher concentrations of Gal80p were 50% more compared to wild-type cells. Cells with higher Gal80p concentration would automatically respond slower compared to cells with lower Gal80p. Overall, these studies indicate that feedback loops of both *GAL80* and *GAL3* are required to maintain cellular homogeneity in terms of the induction state as well as enabling rapid response. Cells disabled for feedback loop of *GAL80* or *GAL3* separately exhibited higher or lower steady-state response to varying galactose concentration as compared to the wild-type cells.

It is clear that the *GAL* regulatory network has the potential for exhibiting multiple steady states by virtue of having feedback loops, but how do we determine the biological relevance of these features? Yeast cells were genetically engineered to express Gal3p or Gal80p from constitutive promoter at a levels equivalent to that of wild-type cell. These cells were grown in galactose (galactose history) or raffinose (raffinose history) for 10 h, subsequently grown in different concentration of galactose for 27 h following which GFP expressed form *GAL* promoter was analyzed at single-cell level. With this experimental regimen it was possible to evaluate the influence of pre-growth conditions on their subsequent behavior, that is, whether the cells can “remember” the past. Wild-type cells pre-grown in galactose responded better at galactose concentrations ranging from 0.02 to 0.15% compared to cells pre-grown in raffinose. As the galactose concentration increased, behavior of the cells coalesced regardless of the pre-growth condition. These results suggest that the cells’ behavior is dependent on the pre-growth conditions.

No difference was observed between raffinose and galactose pre-grown cultures of cells expressing *GAL3* independent of positive feedback loop, unlike what occurs in a wild-type cell. This indicates that the positive feedback loop of *GAL3* makes the cell “remember” the past growth condition. If Gal80p is constitutively expressed, the range of galactose concentration over which the system displayed persistent memory was significantly widened compared to the wild-type. That is, removal of the *GAL80* feedback loop with intact positive feedback loop of *GAL3*, accentuates the “memory phenomenon”, implying that the *GAL80* negative feedback loop counteracts the *GAL3* positive feedback loop. If so, what purpose does the negative feedback loop of *GAL80* serve? Detailed analysis indicated that the negative feedback loop architecture prevents the cells from being trapped in phenotypic states that may not be optimal. The interplay between these loops therefore assists the cells of a population to exist in wide spectrum of expression states. This is an extremely important attribute for microorganisms, which probably allows them to scan a whole range of environmental conditions. That is, phenotypic diversity can be instituted in cells that are otherwise genetically identical.

References

Becskel MA, Oudenaarden A (2005) Enhancement of cellular memory by reducing stochastic transitions. Nature 435:228–232

Bhat PJ, Venkatesh KV (2005) Stochastic variation in the concentration of the repressor activates *GAL* genetic switch: implications in the evolution of regulatory network. FEBS Lett 579:597–603

Goldberger RF (1974) Autogenous regulation of gene expression. Science 183:810–816

Hawkins MK, Smolke CD (2006) The regulatory roles of the galactose permease and kinase in the induction response of the *GAL* network in *Saccharomyces cerevisiae*. J Biol Chem 281:13485–13492

Holde KE, Johnson van WC, Ho PS (1998) Principles of physical biochemistry. Prentice Hall, Upper Saddle River, NJ

Hwang D, Smith JJ, Leslie DM, Weston AD, Rust AG, Ramsey S, Atauri P, Siegel AF, Bolouri H, Aitchison JD, Hood L (2005) A data-integration methodology for systems biology: experimental verification. Proc Nat Acad Sci USA 102:17302–17307

Kew OM, Douglas HC (1976) Genetic co-regulation of galactose and melibiose utilization in *Saccharomyces*. J Bacteriol 125:33–41

McAdams HH, Adam A (1999) It is a noisy business. Trends Genet 15:65–69

Perutz MF (1998) I wish I had made you angry earlier. Oxford University Press.

Ptashne M (1992) A genetic switch: phage λ and higher organisms. Cell Press and Blackwell Scientific Publications, Cambridge

Ramsey SA, Smith JJ, Orrel D, Marelli M, Peterson TW, de Atauri, P Bolouri H, Aitchison JD (2006) Dual feedback loops in the *GAL* regulon suppress cellular heterogeneity in yeast (2006). Nature Genet 38:1082–1086

Verma M, Bhat PJ, Venkatesh KV (2003) Quantitative analysis of *GAL* genetic switch of *Saccharomyces cerevisiae* reveals that nucleocytoplasmic shuttling of Gal80p results in a highly sensitive response to galactose. J Biol Chem 278:48764–48769

8.3 Galactose Metabolism and Evolution

8.3.1 Introduction

In addition to serving as components of cellular fabric and as source of chemical energy, carbohydrates have important biological roles in processes as diverse as cell–cell recognition and signal transduction. Carbohydrates impart enormous functional diversity because of their ability to form not only linear but also branched polymers with enormous structural variation. The transition from single cell to multicellularity obviously had a requirement for structurally diverse molecules with unique specificity, which was fulfilled by the carbohydrates. No wonder that cells of multicellular organisms are decorated by the carbohydrate moieties of glycoconjugates. The synthesis of these complex molecules is not guided by any known code. Therefore, the trajectory of the structural and functional evolution of carbohydrates has been difficult to decipher. Because of these reasons, our current understanding of how cells have evolved to recruit carbohydrates for varied functions is limited. Evolution of galactose metabolism and its regulation poses interesting puzzles at biochemical, functional, and regulatory levels, and provides a unique opportunity to study the evolutionary forces that shaped galactose metabolism in different organisms.

8.3.2 Evolution of Galactose Metabolism

Free galactose is not an abundant sugar in nature, and in this respect, it is an unconventional carbon source. Nevertheless, yeast has developed a sophisticated and highly inducible catabolic pathway. Galactose mainly exists as a disaccharide with glucose either as melibiose or as lactose of plant and animal origin, respectively. Glucose is the precursor of galactose. For heterotrophs like yeast or humans, galactose serves as the carbon source by getting converted to glucose. Galactose and glucose therefore are freely interconverted and the central feature of this is epimerization. While such reactions are quite common, recruitment of three enzymes solely for converting galactose to glucose during its catabolism is a metabolic strategy unique to galactose metabolism. For example, of the three enzymes required for converting mannose to glucose, only one enzyme is unique to this pathway, while the other two are members of glycolysis.

Synthesis of galactose from glucose is not the exact reversal of the Leloir pathway, but uses epimerase catalyzed reactions in the reverse direction. Phosphoglucomutase converts glucose-6-phosphate to glucose-1-phosphate, which is acted upon by UDPglucose/galactose pyrophosphorylase to form UDPglucose, the first nucleotide sugar to be discovered by Leloir during the elucidation of the galactose catabolic pathway. UDP glucose is then epimerized to UDPgalatcose, which is converted to galactose −1-phosphate by UDP glucose/galactose pyrophosphorylase. The conversion of galactose1-phosphate to glucose 1-phosphate can also occur in the reverse direction, resulting in the catabolism of galactose (Fig. 8.3.1). However, this pathway normally does not operate for catabolic purposes due to the low levels of UDPglucose/galactose phosphorylase. Accordingly, it has been demonstrated that over-expression of UDP glucose/galactose pyrophosphorylase overcomes the catabolic block imposed by *gal7* but not *gal10* mutation.

Galactose Utilization: In bacteria, in addition to the carbon and energy source, galactose is a part of the receptor required for phage recognition. As expected, epimerase minus *Salmonella* mutants, are defective in galactose utilization. They are also resistant to infection by phage P-22, due to their inability to synthesize galactose from glucose. It is a paradox that despite an apparent advantage of a mutation in epimerase, which confers resistance to phage infection, bacteria seem to have retained the pathway for galactose synthesis. It has been argued that the

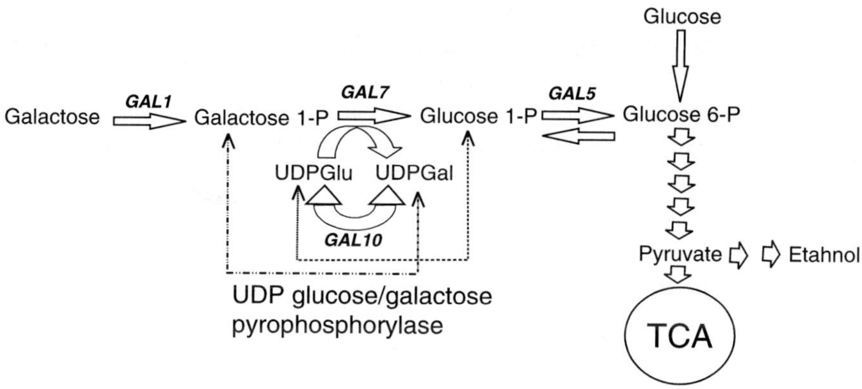

Fig. 8.3.1 Galactose metabolic pathway

phage infection provides a mechanism to acquire new genes through transduction, the advantage of which probably outweighs the disadvantage of maintaining a metabolic pathway.

The bloodstream form of *Trypanosoma brucei*, a protozoan parasite, contains 10^6 variable surface glycoprotein molecules per cell, which are rich in galactose. This parasite lacks the ability to take up external galactose and appears to be devoid of a galactose catabolic pathway, but it has retained galactose 4-epimerase, which is essential for its survival. That is, trypanosoma synthesizes galactose from glucose through the endogenous pathway described above. It lost the catabolic pathway probably to ensure that galactose is exclusively available for using it as a chemical tag. It appears that the demand for galactose as a chemical tag far outweighs its need as a carbon source.

Unlike the two examples above, in yeast, galactose is used only as a source of carbon and energy. That is, strains disrupted for Leloir enzymes grow normally on carbon sources other than galactose and no unique phenotype has been observed with yeast lacking galactose catabolic pathway. Recently, a few species of yeast have shown to be devoid of whole galactose metabolic pathway (see below), further supporting the dispensability of the galactose metabolic pathway in yeast.

In humans, galactose is not an essential dietary component as they can synthesize galactose endogenously from glucose as mentioned before. Humans homozygous for galactokinase and transferase deficiency occur at a frequency of 1:30000, but humans homozygous for the complete absence of epimerase are not present. This argues that in humans, the endogenous pathway of galactose synthesis is essential for viability, as in the case of trypanosoma, but unlike trypanosoma, the galactose catabolic pathway is conserved, and seems to be active especially during infancy. In fact, post-natal development is closely associated with the utilization of galactose, mostly available as lactose present in milk.

Due to the presence of the endogenous pathway of galactose synthesis, individuals homozygous for transferase deficiency suffer from a plethora of physiological disturbances, even if galactose is withheld from the diet. This is due to the accumulation of galactose-1-phosphate originating from endogenous galactose. The underlying metabolic causes of galactose-1-phospahte toxicity observed from bacteria to humans is not clearly understood. Surprisingly, mice deficient for transferase accumulate as much galactose-1-phosphate as transferase deficient patients but do not exhibit any known symptoms of galactose toxicity.

8.3.3 Evolution of Genomic Organization of Galactose Metabolic Enzymes

Evolution of genetic regulation of galactose metabolism has been studied in great detail in bacteria and yeast. The regulation differs in mechanistic details, but exhibits a common regulatory theme: that is, the presence of both positive and negative regulation. A noteworthy difference between bacteria and yeast is in the recruitment of structural genes for galactose utilization. The natural habitat of *E. coli* is the gut, while that of *K. lactis*, a not-so-distant cousin of *S. cerevisiae*, is milk, and it is often referred to as "milk yeast". Lactose, an abundant disaccharide present in milk, upon hydrolysis by β-galactosidase produces galactose and glucose, which serves as a source of carbon and energy. In *E. coli*, β-galactosidase is a member of *lac* operon, which is induced by allolactose, a derivative of lactose. *E. coli* has a separate *GAL* operon induced by galactose. In *K. lactis*, β-galactosidase is a member of *GAL* regulon and is induced by galactose and not by lactose or its derivative. Similarly, *S. cerevisiae*, hydrolyses melibiose present in fruits through α-galactosidase to obtain glucose and galactose as carbon and energy. α-galactosidase is a member of the *GAL* regulon of *S. cerevisiae*, like β-galactosidase, of *K. lactis*. It is quite clear that during evolution, yeast seems to have preferred to do away with a separate regulon for regulating β-galactosidase or α-galactosidase. It is not clear whether this difference between yeast and *E. coli* reflects in someway the symbiotic life of *E. coli* from the saprophytic life of yeast or has to do with other contextual considerations.

Functionally related genes are clustered on the chromosome more often than the functionally unrelated genes, suggesting that genome organization is not driven by random events without any regard to their functional role. For example, in prokaryotes, functionally related genes exist as operons. In yeast, *GAL* genes are clustered even in distantly related species, suggesting a strong selective pressure to maintain the genome organization of *GAL* genes. Recently it has been shown that in *S. naganishii*, a close relative of *S. cerevisiae*, the *GAL10* gene is not clustered with *GAL7* and *GAL1* (Fig. 8.3.2), but is instead located 10 kb away from the *GAL7-1* locus on the same chromosome. The teleological reason for this translocation of *GAL10* is not clear. It is translocated probably to acquire additional regulatory features for expression. For example, in *E. coli, GAL* operon is controlled by two promoters, P1 and P2. There is ample experimental evidence to suggest that the P2 promoter primarily involved in basal the expression of pro-

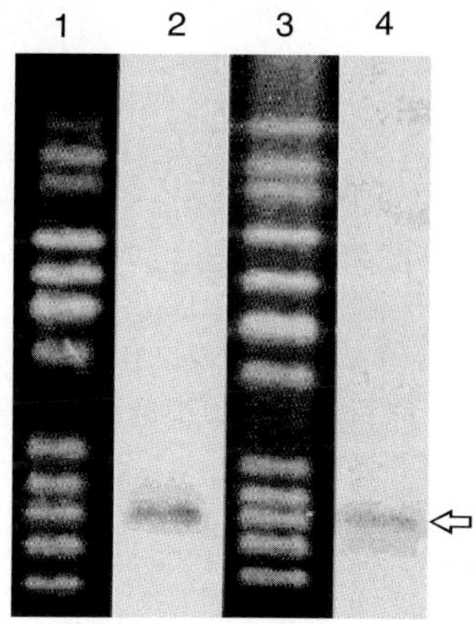

Fig. 8.3.2 Chromosomal localization of *GAL* cluster in *Saccharomyces naganashii*: Chromosomes of *S. naganashii* were separated by pulsed-field gel electrophoresis in duplicate. One set was stained with ethidium bromide (*lanes 1 and 3*), the other was transferred to membranes probed with GAL1-7 (*lane 2*) or GAL10 (*lane 4*). The band picked up by GAL1-7 or 10 probe (indicated by arrow) corresponds to chromosome XI (reproduced with permission from Kodama 2003)

moter proximal epimerase required for the anabolic purposes while P1 promoter turns on the transcription of the whole operon consisting of epimerase galactokinase and transferase. If, for example, a similar need also exists in *S. naganishii*, epimerase would have got translocated to acquire additional regulatory features for basal and induced expression. As discussed before, *GAL10* is required for both anabolic and catabolic purposes, unlike galactokinase and transferase, which are required only for catabolism.

Yeast epimerase is quite unique in another way. It is about twice the size of bacterial or human epimerase. The N-terminal 1-377 residues show significant similarity to the epimerase of bacterial and human epimerase while the 377–699 shows extensive identity to bacterial or human mutarotase. Using classical protein purification techniques, it has been demonstrated that *S. cerevisiae* epimerase indeed contains mutarotase activity. In *E. coli*, the mutarotase is separately encoded by *gal*M and is the fourth member of the *GAL* operon. In humans, it is encoded by an independent genetic locus.

8.3.4 Adaptive Evolution of Galactose Metabolism

As discussed above, galactose metabolic pathway has been molded in different ways in different organisms by evolutionary forces. One of the most challenging tasks in evolutionary biology is to relate and understand how ecological or other unknown forces mold the genome. This challenge becomes even more complicated especially when unrelated organisms are considered for the analysis. Availability of a large number of closely related yeast genome sequences provided a unique opportunity to address the above issues with respect to galactose metabolism (Fig. 8.3.3). Of the 11 closely related yeast species, four of them cannot metabolize galactose.

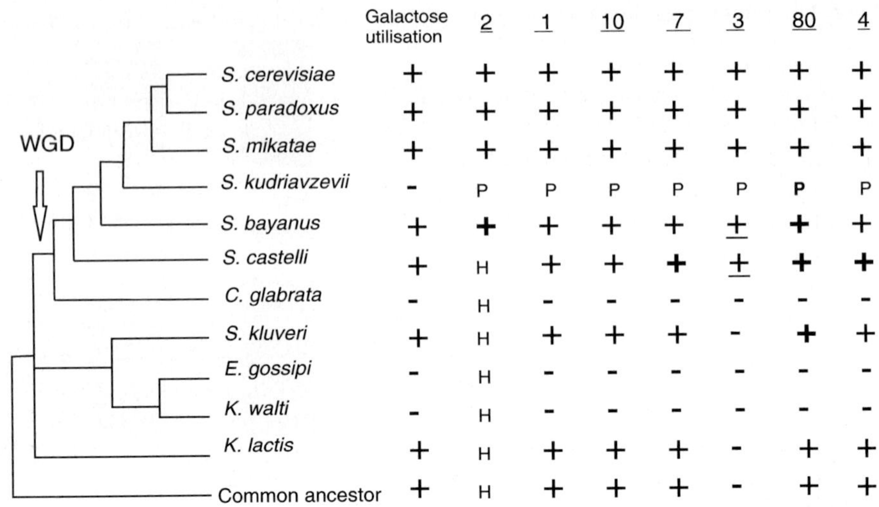

Fig. 8.3.3 Evolution of galactose metabolic pathway in various yeast species. Ability and inability to use galactose is indicated by + or –, respectively. The members of the *GAL* genes are indicated by their numbers. Absence and presence of genes is indicated by + or –. *P* represents pseudogenes. *H* represents members of the hexose transport family, could be orthologues of *GAL2*. *Arrow* indicates the whole genome duplication (*WGD*) that occurred during the evolutionary branching (adapted with permission from Hittinger et al. 2004)

Of the four that cannot utilize galactose, three species broke off quite early from the evolutionary branch. In *S. kudriavzeviii*, more closely related to yeast *cerevisiae, GAL* genes are present as pseudogenes. Unlike *S. cerevisiae*, the natural habitat of *S. kudriavzevii* is soil. It appears that this change in lifestyle did not put selective pressure for retaining the metabolic pathway. Given the close phylogenetic relationship with *S. cerevisiae*, and the fact that the *GAL* genes are not vanished from the genome of *S. kudriavzeviii*, the loss of galactose metabolic pathway appears quite recent.

In *Saccharomyces cerevisiae*, Gal4p also regulates genes that are peripherally involved in the galactose metabolic pathway such as *MTH1, PLC10 GCY1*. These genes have not been lost in *S. kudriavzevii*, suggesting that the loss is specific to *GAL* pathway. Interestingly, the upstream sequences of the *GCY1, MTH1, PLC10*, which are not directly related to galactose metabolism, have retained orthologous Gal4p-binding sequence despite the loss of *GAL4*. Another intriguing observation is that in *E. gossipi* although *GAL* pathway has been lost, the presence of a syntanic *GAL4* ortholog suggests that it has been retained for another function. The above results suggest that it is possible that genes such as *GCY1, MTH1, PLC10* and orthologue of *GAL4*, might have a more pliotropic role and have been therefore retained. This is a fascinating example of evolution in action.

8.3.5 Evolution of Regulatory Network of Galactose Metabolism

K. lactis is evolutionarily separated from *S. cerevisiae* 1.5×10^8 years ago. The natural habitat of *S. cerevisiae* is fruits, while that of *K. lactis* is milk. These two yeasts are similar with respect to many aspects of galactose metabolism but differ in the fine regulation of galactose genetic switch. First, unlike *S. cerevisiae, K. lactis* has a significant basal expression of all the structural genes including the *GAL* cluster. Second, *K. lactis* lacks *GAL3*, the paralogue of *GAL1*. Accordingly, the signal-transduction function in *K. lactis* is carried out by the GAL1-encoded galactokinase, which is a bifunctional protein like the yeast Gal1p. Third, in *K. lactis* it has been shown that the Gal1p enters the nucleus to inactivate Gal80p. In *S. cerevisiae*, experimental evidence suggests that Gal3p is exclusively cytoplasmic and Gal80p is a nucleocytoplasmic shuttle protein. As mentioned before, the induction is brought about by the sequesteration of Gal80p in the cytoplasm. Lastly, in *K. lactis, LAC9* functional counterpart of the *GAL4*, is under autoregulation, unlike *S. cerevisiae*. Whether these differences in regulation reflect a corresponding change in their lifestyle under the natural condition is yet to be explored.

In addition to the differences at the level of regulation, changes have occurred at the level of protein structure function relationship. While Sc *GAL1/GAL3* share amino-acid sequence similarity with *KlGALl*, neither *ScGAL1* nor *GAL3* complements Kl *GAL1* defect. However, if Kl *GAL80* is replaced by Sc*GAL80* in *K. lactis*, then

Box 8.3.1 *Pulsed-field gel electrophoresis*

In conventional agarose gel electrophoresis, small DNA fragments pass through the pores of the gel by a sieving effect. In DNA, the charge-by-mass ratio is invariant of the molecular weight, and therefore the separation is a function of molecular weight. Above a certain threshold of size, DNA molecules attain a compact form too large to enter the pores. Often, when plasmids are isolated from *E. coli*, it can be seen that high-molecular-weight DNA remains at the origin. Such forms can only enter the gel if they can attain an extended form with the leading end migrating into the gel. To achieve separation of these large fragments, pulsed-field gel electrophoresis (a modified form of agarose gel electrophoresis) is used. This technique works on the principle that under a discontinuous electric field, DNA molecules are periodically forced to change their conformation and direction. The time taken to reorient the DNA molecule in the new electric field is a function of its size. This allows the separation of DNA fragments up to several megabases in size. The discontinuous electric field is generated by alternatively activating two differentially oriented fields or a single electric filed with a periodic reversal of the field. The difference between the conventional and pulsed-field electrophoresis is somewhat similar to a 100-m sprint and an obstacle course.

Box 8.3.2 *Syntany*

If a pair of genes mapping to a chromosome in one species also maps to a chromosome in another species, then the pair of genes is considered syntanic. That is, there are regions or segments containing matching genes across two species. Several such syntanic groups are identified in *A. gossipi* and *K. walti* with reference to *S. cerevisiae.* In the majority of these cases, the homology relation of single genes or a subgroup of genes alternated between two *S. cerevisiae* chromosomes, and this is referred to as double conserved syntany. If the orientation and the distance between the two genes are preserved, then it reflects the configuration of the genes in the progenitor of the two analyzed species. If the gene order of the syntanic pair is conserved, it is called a "linkage conservation". A 1:1 mapping between syntanic region is observed between close relatives. Extensive double syntany is taken as proof that *S. cerevisiae* genome got duplicated after the *Kluyveromyces* and *Saccharomyces* genus got separated.

the Sc Gal1p and Gal3p transduce the signal. These results indicate that *S. cerevisiae* signal transducers do not recognize KlGal80p. On the other hand, Kl *GAL1* can suppress the long-term adaptation of *S. cerevisiae*, indicating that KlGal1p recognizes ScGal80p. Unlike the signal transducers, Sc *GAL80* and *GAL4* complement the corresponding defect in *K. lactis* and vice versa.

8.3.6 Genome Duplication in Saccharomyces

Genome duplication is one of the principal forces that drive the evolution of protein function. What is the fate of duplicated genes? After duplication, two genes with identical function cannot be maintained unless there is a selective pressure for over-expression. If not, duplicates are either deleted, modified to perform related or totally unrelated function. It has been proposed that both the paralogues will be retained provided each daughter gene adapts part of the function of the parental gene. This sub-functionalization can occur due to changes in the expression pattern or due to changes in protein function. *GAL3* is a classic example where it has retained a part of the parental function, that is, signal transduction, but lost the expression pattern as well as the galactokinase function. The purpose of this subfunctionalization is not apparent but it is possible that by recruiting a dedicated signal transducer, it could abolish the basal expression of the Leloir enzymes.

It is predicted that the whole genome duplication of *Saccharomyces* occurred after *K. lactis, K. walti* and *A. gossipi* diverged from *Saccharomyces* more than 100 million years ago. After the duplication, most of the genes were deleted and gene order was rearranged by reciprocal translocation. There are at least 145 sister regions in *S. cerevisiae* covering 88% of the genome. At least 457 gene pairs arose due to whole genome duplication. The rest of the paralogues are thought to have evolved by local duplication events. The most significant differences between *Kluyveromyces* and *Saccharomyces* is that the former is an aerobe and the latter is a facultative anaerobe. The second difference is that *Saccharomyces* is petite-positive while *Kluyveromyces* is petite negative. It has been proposed that whole genome duplication and extensive genomic rearrangement would have resulted in the differences in the phenotypes.

References

Anders A, Liele H, Franke K, Kapp L, Stelling J, Giles ED, Breunig KD (2006) The galactose switch in *K. lactis* depends on nuclear competition between GAL1 and GAL4 for GAL80 binding. J Biol Chem 281:29337–29348

Bhat PJ, Murthy TVS (2001) Transcriptional control of GAL/MEL regulon of yeast *Saccharomyces cerevisiae*: mechanisms of galactose expression signal transduction. Microbiology 40: 1059–1066

Bhat PJ (2003) Galactose-1-phosphate is a regulator of inositol monophosphatase; A fact or a fiction. Medical Hypothesis 60:123–128

Dietrich et al (2004) The *Ashbya gossypii* genome as a tool for mapping the ancient *Saccharomyces cerevisiae* genome. Science 304:5668–5674

Frey PA (1996) The Leloir pathway: a mechanistic imperative for three enzymes to change the stereochemical configuration of a single carbon in galactose. FASEB J 462:461–470

Hittinger CT, Rokas A, Carroll SB (2004) Parallel inactivation of multiple *GAL* pathway genes and ecological diversification in yeast. Proc Nat Acad Sci USA 101(9):14144–14149

Hittinger CT, Carrol SB (2007) Gene duplication and the adaptive evolution of a classic genetic switch. Nature 449:677–681

Janine R, Maria R, Guther LS, Miline KG, Ferguson MAJ (2002) Galactose metabolism is essential for the African sleeping sickness parasite *Trypanosoma Bruci*. Proc Nat Acad Sci USA 99:5884–5889

Kalckar HM (1966) High-energy phosphate bonds: optional or obligatory? In: Cairns J, Stent G, Watson JD (eds) Phage and origins of molecular biology. Cold Spring Harbor Laboratory, Long Island, NY, pp 43–49

Kodama T, Hisatomi T, Kakiuchi M, Aya R, Yoshida K, Bando Y, Takami T, Tsuboi M (2003) Unique distribution of *GAL* genes on chromosome IX in the yeast *Saccharomyces naganishi*. Curr Microbiol 47:497–500

Kellis M, Birren W, Lander ES (2004) Proof and evolutionary analysis of ancient genome duplication in the yeast *Saccharomyces cerevisiae*. Nature 428:617–623

Lai K, Elsas LJ (2000) Over-expression of human UDPglucose pyrophosphorylase rescues galactose-1-phosphate uridyltransferase deficient. Biochem Biophys Res Commun 271:392–400

Mehta DV, Kabir A, Bhat PJ (1999) Expression of human inositol monophosphatase suppresses galactose toxicity in *Saccharomyces cerevisiae*: possible implications for galactosemia. Biochem Biophys Acta 1454:217–226

Moller K, Olsson L, Piskur J (2001) Ability for anaerobic growth is not sufficient for the development of the petite phenotype in *Saccharomyces kluyveri*. J Bacteriol 183:2485–2489

Rubio-Taxiera M (2005) A comparative analysis of the *GAL* genetic switch between not-so-distant cousins: *Saccharomyces cerevisiae* and *Kluyveromyces lactis*. FEMS Yeast Res 5:1115–1128

Semsey S, Vernik, K, Adya S (2006) Three-stage regulation of the amphibolic *GAL* regulon: from repressosome to GALR-free DNA. J Mol Biol 358:355–365

Teichmann SA, Veitia RA (2004) Genes encoding subunits of stable complexes are clustered on the yeast chromosomes: an interpretation from a dosage-balance perspective. Genetics 167:2121–2125

8.4 *GAL* Switch as a Tool

8.4.1 Introduction

It is clear from the foregoing that the lessons learned from the *GAL* system have had far-reaching implications beyond the field of regulation of gene expression. I have already alluded to its paradigmatic use for the in-depth understanding of biological themes. More recently, this system has been used as a test-bed to evaluate, validate, and develop methods for data integration for systems biology, for high resolution mapping of genomic mutations that occur during adaptive evolution as well as to correlate the nonlinear relationship between genotype and the phenotype. This system is evolving continuously to keep pace with developments and preparing itself for futuristic needs. However, the most important application of the *GAL* system thus far is its use as a regulatable genetic switch. Another major fallout of the basic understanding of this switch is the development of a technique for identifying in vivo protein–protein interactions. In this section, I shall focus on these two topics although many variations of these basic techniques have been developed.

8.4.2 High-level Protein Expression

Obtaining a large amount of proteins in pure form is a bottleneck in basic and applied field. Despite the advances in recombinant DNA technology, protein expression has remained a technical challenge. No wonder then a large number of expression systems of prokaryotic and eucaryotic origin have been developed. Yeast is one of the attractive hosts for the production of heterologous proteins owing to the fact that many aspects of yeast biology are well understood. We have already encountered an example of over-expression of Gal4p driven by the *GAL1/10* promoter (see Sect. 5.2).

In general, the desired ORF to be expressed is cloned under the control of *GAL1/10* promoter. This expression cassette can either be maintained as an extra chromosomal element in high copy number or can be integrated into the chromosome for stable maintenance. The recombinant strains are grown under appropriate conditions to high cell density before the promoter is turned on by the addition of galactose. This approach bypasses the toxic affects that the heterologous protein may have on cell growth.

One of the limitations of this system is the low abundance of Gal4p, which limits the expression of the desired protein from the *GAL* promoter. This limitation is overcome by constructing yeast strains wherein “GAL10:GAL4” expression cassette, (that is GAL4 expression is under the control of GAL10 promoter) is integrated into the yeast genome. This allows the over-expression of Gal4p upon the addition of galactose, which in turn drives the expression of the desired protein from the *GAL* promoter. A second limitation is that galactose, the inducer, is also a carbon source and is depleted from the medium as the cells grow. To circumvent this problem, yeast strains have been engineered that lack galactose metabolic pathway by disrupting galactokinase, the first enzyme in galactose metabolism. Such a strain can be grown in carbon sources such as ethanol and the *GAL* promoter can be induced at a much lower concentration of galactose as it is not metabolized.

8.4.3 Dihybrid Analysis

Over the years, significant effort has been directed to develop techniques to detect and analyze protein–protein interactions. This is necessitated by the sensitivity and diversity of protein–protein interactions under cellular conditions. An impressive array of genetic, biochemical, and biophysical approaches to study protein–protein interactions have been developed. Despite this, a generic approach linking in vivo protein–protein interaction to a biological readout was not available. Dihybrid analysis filled this void and subsequently it has been modified in different ways to suit the specific demands of researchers. Here, I shall focus on the basic principles with some examples to illustrate its varied use in (a) testing the in vivo interactions between any two proteins, (b) in isolating the interacting protein partner of a target protein, and (c) in conducting genetic analysis.

The idea of dihybrid analysis emerged from many independent observations made during the analysis of Gal4p. Two key experiments are worth recalling in this regard. First, DNA binding and the transcriptional activating domains of Gal4p can be separated and grafted to other proteins without the loss of the biological activity. Second, hybrid Gal80p, consisting of transcription-activating domain of Gal4p, activated the *GAL* system in the absence of galactose in *gal80* strain carrying a *GAL4* allele, which is capable of interacting with Gal80p but is defective in transcription activation. This implied that recruitment of transcriptional activation domain through non-covalent interaction between Gal4p and Gal80p to the proper site is sufficient for transcriptional induction.

Proof of principle of dihybrid analysis: In vitro data had suggested that Snf4p associates with Snf1p, a serine-threonine-specific protein kinase (see Chap. 7.2). This interaction was used to demonstrate the concept of dihybrid analysis. Hybrid proteins consisting of N-terminal 1–147 amino acids of Gal4p fused in frame to the entire Snf1 protein was cloned under the control of *ADH1* promoter in a single-copy plasmid bearing *URA3* marker. Snf4p protein fused in frame to the Gal4p transcriptional activation domain II was cloned under the control of *SNF4* promoter in a mulitcopy plasmid with *LEU2* marker. A *gal4 gal80* strain bearing a *GAL1/10* β-galactosidase expression cassette integrated into the chromosome was transformed with the above expression plasmids. Transformants bearing both the plasmids expressed β-galactosidase (Fig. 8.4.1). Neither of the plasmids alone activated the β-galactosidase expression, which demonstrated that the transcriptional activation was reconstituted due to the interaction between Snf1p and Snf4p.

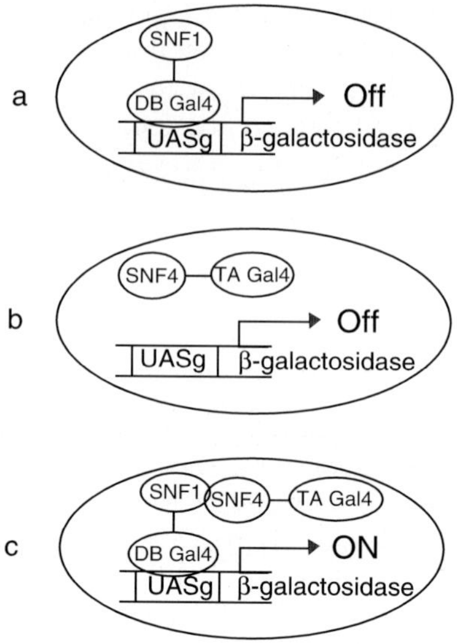

Fig. 8.4.1 Proof of principle of dihybrid analysis. A *gal4gal80* strain containing an integrated β-galactosidase expression cassette driven by the *GAL1* promoter was transformed with indicated fusion plasmid. Strain transformed with bait plasmid consisting of a fusion of Gal4p DNA-binding domain with SNF1 **a** or the prey plasmid consisting of fusion with the transcription domain of Gal4p with SNF4 **b** does not express β-galactosidase. Strain consisting of both the bait and prey plasmid **c** expresses β-galactosidase, indicating that Snf1p and Snf4p interact

8.4.4 Dihybrid Approach for Genetic Analysis

Although this technique was originally developed for detecting protein–protein interaction, it has been used for various other purposes. I have discussed below a few examples that illustrate the versatility of this technique.

Gene isolation: Dihybrid technique was quickly adapted to screen libraries for the isolation of genes encoding proteins that interact with the target protein. The success of this approach was demonstrated by isolating interacting protein of Snf1p kinase. A *gal4gal80* mutant yeast strain bearing an integrated copy of *GAL1*:β-galactosidase as a reporter was used as a recipient for screening the yeast genomic library. The above strain was transformed with the plasmid expressing a hybrid protein consisting of *GAL4(1-147)* fused with *SNF1* gene. This transformant was subsequently transformed with a plasmid library consisting of yeast genomic fragments fused to transcriptional activation domain II of Gal4p. Double transformants selected based on the auxotropic marker were screened for β-galactosidase expression in the subsequent step. Transformants that expressed β-galactosidase are expected to contain fragments of the yeast genome encoding proteins that interact with Snf1p (Fig. 8.4.2). This screen picked up four independent clones referred to as SIP (Snf1p interacting

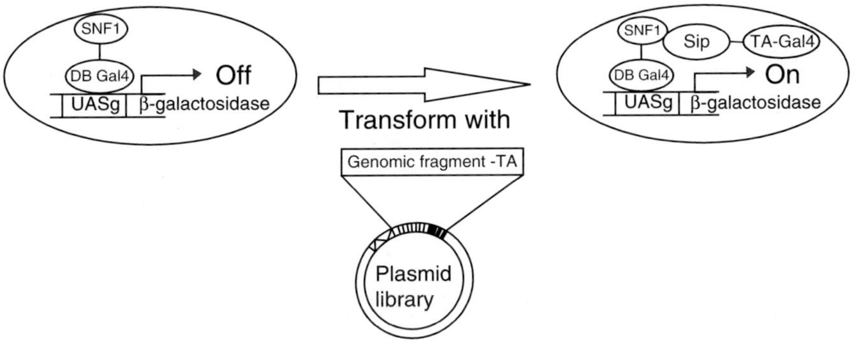

Fig. 8.4.2 Isolation of gene using dihybrid analysis. A *gal4gal80* transformant bearing the bait plasmid containing an integrated β-galactosidase expression cassette driven by the *GAL1* promoter was transformed with a library of recombinant plasmids containing the genomic fragments fused in frame with the transcription activation domain. Transformants initially picked up based on the auxotrophic markers were then screened for the expression of β-galactosidase. Plasmids are isolated from transformants expressing β-galactosidase and analyzed

proteins), one of which was later shown to be a component of *SNF1* kinase and further analysis showed it is a member of the protein scaffold.

Genetic analysis: Whether Gal80p dissociates or remains bound to Gal4p upon induction is not clear. The fate of Gal80p-Gal4p interaction upon induction was tested even before it was known that Gal3p and Gal80p interact to transducer the signal. To test this, a yeast strain disrupted for *GAL80*, but bearing a transcriptionally inactive Gal4p, which can still interact with Gal80p was constructed. A plasmid that codes for a wild-type Gal80p (control) or a hybrid Gal80p consisting of transcriptional activation domain of a viral transcription activator, VP16, was introduced into the above strain. If upon galactose addition GAL80-VP16 dissociates, then the strain is not expected to induce β-galactosidase, since the Gal4p is unable to activate the transcription. On the other hand, if it does not dissociate, β-galactosidase would continue to express regardless of the system is induced or not. Upon galactose addition, β-galactosidase expression was observed (Fig. 8.4.3). This lead to the idea that Gal80p may not dissociate from Gal4p, even upon induction.

Dihybrid analysis was also recently used to isolate substitution mutations in Gal3p that would interact with Gal80p even in the absence of galactose. Here, Gal3p was expressed as a fusion protein consisting of Gal4pDNA-binding protein (Bait). If galactose is not provided, this fusion protein cannot induce expression of a reporter gene fostered by *GAL* promoter in *gal4gal80* strain co-transformed by a plasmid-expressing Gal80p-VP16 fusion protein (Prey). *GAL3* present in the bait plasmid was randomly mutagenized in specific domains using a PCR mutagenesis scheme and the yeast strain was co-transformed with prey plasmid followed by screening the double transformant that expresses the reporter gene in the absence of galactose. The reasoning is that if a constitutive mutation is introduced into Gal3p, then it

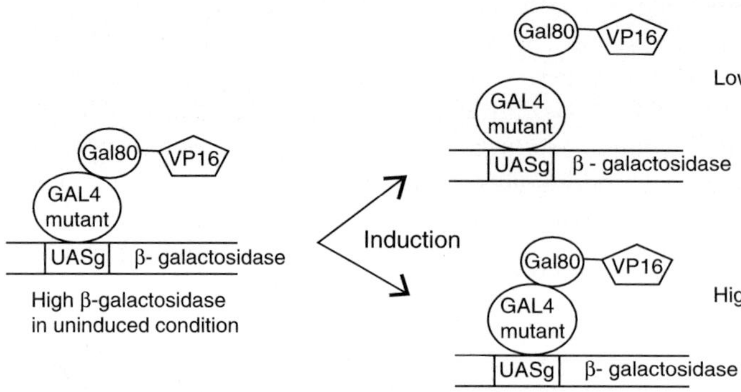

Fig. 8.4.3 Dihybrid analysis of Gal80p–Gal4p interaction

should interact with the Gal80p-VP16 to activate the transcription of the reporter, even if galactose is absent. The bait plasmid was isolated from the transformant that expresses the reporter constitutively and analyzed further. A similar strategy was adapted to isolate domain-specific mutations in Gal80p (see Chap. 6.2).

8.4.5 Genome-Wide Protein–Protein Interaction

The power of dihybrid analysis unfolded after the whole genome sequence of many organisms became available. This technique could be applied to include all the possible permutation and combination of interactions between the yeast proteins. For example, in yeast, a total of 4×10^7 combinations of interactions are possible. To determine protein–protein interaction at the genomic level, individual ORFs were separately amplified and cloned into AD (bait) and TA (prey) plasmids. These recombinant plasmids express every ORF as a fusion protein with DNA-binding domain as well as transcription domain, respectively. Haploid yeast strains of both mating type with appropriate auxotropic marker were individually transformed with bait and prey plasmids. Transformants bearing the bait and the prey plasmids are mated to generate diploid. In diploids, reporter gene activation occurs only if the two gene products interact. The diploids containing the prey and bait plasmids that encode the interacting proteins are selected or screened based on reporter assay or growth phenotype or both.

A one-by-one matrix approach and a high-throughput matrix approach have been used to determine the protein–protein interaction using the above technology. For example, in one study using a one-by-one matrix approach, 281 interactions have been detected out of a total of 192 bait versus 6,000-prey plasmids tested. In the high-throughput approach, a total of 286 interactions have been detected out of a total of 5,341 bait and prey plasmids tested. The yeast protein interaction database holds many biologically relevant interactions, which testifies to the importance of this approach. Despite many limitations of the Y2H approach, protein–protein interaction maps of several organisms have been developed. It is important to note that this technique relies purely on a physical interaction between two proteins regardless of their biological relevance. Conversely, it is equally possible that the biologically relevant interactions may not be picked up using this technique. One of the limitations is due to the dependence of this approach on transcriptional activation, which in turn relies on many other biological processes. For example, this assay would work only if the biological activity of the individual protein is retained even after fusion.

8.4.6 GAL Switch as a Tool in Higher Organisms

Turning an exogenous genetic circuit on or off in living organism such as drosophila, provides an unprecedented opportunity to study the consequence of overexpression, misexpression, and silencing gene expression on various biological processes. As early as 1988, it was shown that Gal4p expressed from alcohol dehydrogenase promoter in drosophila can activate β-galactosidase cloned under UASg in cells where alcohol dehydrogenase is normally expressed. While subsequent studies showed that Gal4p can function as a transcriptional activator in a variety of organisms including humans, plants and flies.

In *Drosophila*, *GAL4* is mobilized throughout the genome by P-element mediated transposition, thus bringing *GAL4* under the expression of the endogenous enhancer elements. For example, in one line of flies, Gal4p might be expressed from the enhancer which functions only in muscle at a particular stage of development while in another, it might express in ocular tissue. Currently there are at least 1,095 characterized *Drosophila* lines, wherein *GAL4* expression is driven by different enhancer elements. In another set of flies, the gene to be expressed is cloned under the UASg and integrated into the genome. Thus, the GAL4 protein and its target genes are initially separated in two separate fly lines. These two lines are crossed to obtain flies that express the transgene in a Gal4p-dependent manner in the tissue of ones interest. In one of the earliest studies, tetanus toxin was expressed either in muscle or embryonic neuron using the GAL system. This study showed that toxin inhibits synaptic transmission when expressed presynaptically in motor neurons but does interfere with the muscle function if expressed postsynaptically.

While the above system allows the tissue-specific expression of the transgene, it does not give the investigator the control to express the transgene temporally. This barrier is overcome by coexpressing a temperature-sensitive allele of *GAL80* from the ubiquitous tubulin promoter. This allele of *GAL80* represses *GAL4* at 19 °C but at 30 °C Gal80p is inactive and Gal4p-dependent expression of the transgene is ensued. This system can be used to evaluate the effect of the turning on or off of a transgene expression during a specified time point, for example during development, simply by lowering or raising the temperature. In one example, flies with eye-specific

GMR-Gal4 driver in combination with UASg-toxin could not be reared due to developmental lethality. However, in the presence of *GAL80*^{ts} allele, normal members of the adults were recovered with normal eyes at 19 °C. If the flies were shifted to 30 °C at different stages of development adult flies with range of phenotypes. These examples clearly illustrate the use of introducing exogenous genetic circuits for genetic analysis.

References

Fields S, Sternglanz R (1994) The two-hybrid system. Trends Genet 10(8):286–291

Field S, Song Ok-Kyu (1989) Novel genetic system to detect in vivo protein–protein interaction. Nature 340:245–246

Hwang D, Smith JJ, Leslie DM, Weston AD, Rust AG, Ramsey S, Atauri P, Siegel AF, Bolouri H, Aitchison JD, Hood L (2005) A data-integration methodology for systems biology: experimental verification. Proc Nat Acad Sci USA 102:17302–17307

Leuther KK, Johnston SA (1992) Non-dissociation of GAL4 and GAL80 in vivo after galactose induction. Science 256:1333–1335

Legrain P, Selig L (2000) Genome-wide protein interaction maps using two-hybrid system. FEBS Lett 480:32–36

Mylin LM, Hopper JE (1997) Inducible expression cassettes in yeast: GAL4. In: Taun R (ed) Recombinant gene expression protocols. Humana Press, Totowa, NJ, Meth Mol Biol Vol. 62

McGuire SE, Roman, Davis RL (2004) Gene expression systems in *Drosophila*: A synthesis of time and space. Trends Genet 20:384–391

Peccoud J, Veldern KV, Podlich D, Winkler C, Arthus L, Cooper M (2004) The selective values of alleles in a molecular network model are context-dependent. Genetics 166:1715–1725

Sweeney ST, Broadie K, Keane J, Niehemann H, O' Kane CJ (1995) Targeted expression of tetanus toxin light chain in *Drosophila* specifically eliminates and causes behavioral defects. Neuron 14:341–351

Yang XE, Hubbard JA, Carlson M (1992) A protein kinase substrate identified by two-hybrid system. Science 257:680–682

White MA (1996) The yeast two-hybrid system: forward and reverse. Proc Nat Acad Sci USA 93:10001–10003

8.5 Lessons Learned

8.5.1 Introduction

What started out as an intellectual curiosity to understand the phenomenon of enzyme adaptation more than century ago has evolved into an elegant model system to understand various facets of biology at both the experimental and theoretical levels. For historical reasons, interest in this field plummeted during the first half of 20th century, only to resurface in the 1940s. Since then, not more than 100 publications would have appeared on this topic. Nevertheless, its contribution to fundamental and applied biology has been impressive. How did this transition occur from a simple model for understanding enzyme adaptation to a model system of far-reaching implications? The answer perhaps lies in the fact that this system has all the attributes for studying the relationship between the genotype, phenotype, and the environment with metabolism as a focal theme. Come to think of it, understanding the molecular details of the evolutionary trajectory of metabolism is at the heart of mapping genotype to phenotype relationship. This is because the genotype and the phenotype, which are otherwise far removed, are connected through a common thread of metabolism that guides an organism to reproduce itself incessantly at the expense of matter and energy. In this book I have attempted to expound these basic principles in the backdrop of genetics and evolution.

As early as the 1940s, John von Neumann drew an analogy between the ability of the living organisms to reproduce and the functioning of mechanical automata. The automaton, as perceived by Neumann, had two components: the software that encodes the information, and the hardware that processes the information. In today's parlance, the software is represented by the genome, consisting of a stretch of four nucleotides arranged in a way unique to a given species with its ability to code a set of instructions depending upon the species and cellular state. We do not yet completely understand how the software is implemented at the level of protein structure starting from the primary amino-acid sequence. The metabolome, proteome, and the transcriptome collectively represent the hardware. Von Neumann suggested that for a faithful reproduction of the living being, both components are essential, but he gave the hardware a logical priority.

What would happen to this basic capability of reproduction if the software and the hardware were to disconnect? Interestingly, such experiments have been carried out by nature. During differentiation of red blood cells (RBC), the precursor cell loses the DNA, the software, but not the metabolic machinery. Although metabolically competent, RBC lacks the fundamental feature of a living being, that is, it cannot divide. In contrast, a virus infecting bacterial or human cells has the software for self-replication, but not the supporting metabolic machinery. In computer language, viruses contain a program but lack the means of implementing it. Thus, the virus needs the metabolic machinery of its host cell to multiply.

On the other hand, the sperm and the egg by themselves cannot reproduce, despite the fact that they possess the metabolic machinery and the genetic apparatus. Sperm and egg acquire the ability to divide only upon fusion. It is not the diploid status per se that confers the ability to reproduce, since organisms with haploid genomes such as yeast can reproduce asexually. The mere presence of the genome and the metabolic machinery does not guarantee the ability to reproduce. The common notion that a "genetic program", as understood in a loose sense, is clearly insufficient to dictate the cytoplasm to execute the life processes. Rather, genetic material and the cytoplasm collaborate in some unique way to elaborate life processes. Understanding this intricate collaboration is the mainstay of modern biology.

8.5.2 Robustness and Fragility

In contrast to our understanding of the role of genetic programming in biological processes, the contribution of the rest of the cellular milieu in the manifestation of the phenotype is only beginning to emerge. A critical feature of living beings is robustness, without which there would be no life on earth. That is, organisms cannot afford to be too sensitive to the onslaught of physical and chemical perturbations. Robustness is the ability to function normally in the face perturbations. To understand the biological significance of robustness, both the nature of perturbations as well as the attribute that responds robustly against the perturbations have to be defined. For example, some proteins retain their function despite mutational assault. It was observed that 84% of β-lactamase amino acids can be substituted without severely impairing the ability to confer antibiotic resistance. Here, the protein structure is robust and has to do with the way in which the primary amino-acid sequence participates in the final structure, a topic not very clearly understood, as mentioned before. Let us consider another simple example. Naturally occurring *Saccharomyces cerevisiae* is diploid. In case of a mutation in one copy of *GAL1* the mutant can still grow on galactose. Here the diploidy provides the robustness.

How do organisms evolve mechanisms to remain robust to conditions that they possibly have not encountered in the past? Yeast seems to have evolved a mechanism to remain robust against the impending assault of gene deletions. For example, more than 50% of the yeast genes can be deleted without any deleterious consequence on fitness under laboratory conditions. One of the mechanisms of achieving robustness is to evolve degenerate "fail safe" mechanisms. A specific example of robustness because of duplicate gene is the signal-transduction pathway of *GAL* regulon of yeast. A mutation in *GAL3* does not completely knock off its ability to use galactose, but only makes it a bit slow. This is because of the degenerate *GAL1* that performs the *GAL3* function. Consider *K. lactis*, another closely related yeast. In case its signal-transduction function is lost, it cannot utilize galactose, as it does not have a degenerate signaling mechanism. Clearly then, robustness is the key for the evolutionary success of an organism.

In addition to the above, many design principles such as modularity in network and feedback regulation are at work in imparting robustness. A natural fallout of increasing robustness to specific perturbations is that the system remains sensitive or fragile for certain other perturbations. This is because an increase in robustness gives rise to complexity, which inevitably leads to fragility of the system. For example, many of the human diseases are due to haploinsufficiency. Here, inactivation of one of the alleles of a diploid cell results in an insufficient amount of the protein product, which leads to an imbalance in the component concentration of protein complexes. Here, the protein complexes probably meant to provide robust output to specific signals are sensitive or fragile to a moderate decrease in component concentrations. Although it appears anti-intuitive that robustness and fragility coexist, they are inseparable. Evolution seems to have cleverly exploited the trade-off between robustness and fragility to its advantage in many instances (see below). Biological processes as complex as development and differentiation occur with astonishing precision and reliability precisely due to this trade-off.

8.5.3 Stochasticity and Phenotypic Variation

No matter what the underlying molecular mechanisms are, eventual success of an organism depends on the phenotype. Although it is known that duplicate genes provide robustness, interactions among a network of molecular components seems to play far more significant role in implementing physiological robustness to very many changing conditions. It is becoming increasingly clear that the non-genetic diversity, a phenomenon that has a far-reaching implication from evolution to differentiation to disease phenotypes, is due to stochasticity or "noise" inherent to the system [noise or coefficient of variation (η) is statistically defined as the ratio between standard deviation (σ) to the mean (N)]. Stochasticity as an intrinsic mechanism to generate diversity or multiple steady states raises the question as to how evolution has succeeded in maintaining order in the midst of chaos. That is, stochasticity intuitively appears to be detrimental, and contradicts the notion that deterministic outputs as complex as life cannot manifest if the underlying molecular mechanisms are inherently stochastic. However, it really does not come as a surprise, since indecisiveness, that is, "to be or not to be" state has an inherent advantage and in fact, on closer look, characterizes the behavior of living beings. It is now accepted that living systems do not function with clockwork precision, but are comfortable with noisy events. Such a mechanism has been shown to be not only restricted to free-living microbes but also observed in multicellular organisms.

There is growing evidence that regulation of gene expression, fundamental to many of biological processes, is inherently stochastic. Stochastcicity is due to intrinsic and extrinsic reasons. First, the biochemical reactions such as transcription, translation are stochastic due to the low concentration of the reacting species resulting in a noisy pattern of gene expression. Second, the inevitable statistical variation during the portioning of regulatory molecules between the daughter cells as they divide indirectly lead to a variation in the expression of genes. Experimentally, it has been demonstrated that the same cell containing two identical genes under the identical regulatory circuit expresses varying levels of the gene product. Stochasticity in gene expression is believed to be the cause for non-genetic phenotypic variation. If so, do we expect it to be a target for natural selection? It has been observed that of the 15 genes involved in essential biological processes in yeast, 14 genes have high transcription with low translations, a property associated with reduced noise in expression. This lends credence to the idea that evolution has ensured that essential cellular functions cannot be subject to the vagaries of stochasticity. This is because the concentrations of essential proteins are maintained within narrow limits to ensure proper cellular functioning. Conversely, non-essential genes invariably exhibit stochastic expression giving rise to a wide spectrum of phenotypic states. Thus, stochasticity can ensure that a population as a whole can sample out widely fluctuating environment, a property that provides a distinct advantage. Under these conditions, an organism that has a robust "Yes or No" phenotype would clearly be at a disadvantage. The advantage of stochasticity as a principle in evolutionary terms stems from its reversibility, unlike a mutational event, which is normally irreversible.

Stochastic phenotypic switching has been thought to be one of the causes of the persistence of bacterial infections following antibiotic treatment. In an antibiotic-treated population, a small fraction of cells which are resistant, probably because of a difference in the gene expression as compared to the majority of the members of the population, survive for an extended period of time. Upon removal of the antibiotic, this small population can emerge and re-establish the infection. However, it is yet to be established that this phenotypic switching that is resistant to antibiotic is due to a stochastic transition in expression of protein. Here, genetic constitution of the individual remains unaltered and yet the phenotype keeps varying. An increase in stochasticity, causing transition from one phenotypic state to another, has implications in the onset of disease states. As mentioned before, haploinsufficiency is a common cause of many diseases. Based on theoretical considerations it has been suggested that haploinsufficient cells are more prone for stochastic interruptions due to increased noise. It was observed that normal and haploinsufficient melanocytes for NF1 gene product exhibit difference in the morphology of dendrites in terms of the length, angle, and number. A mathematical model that incorporated a term for noise could adequately explain these differences, suggesting the role of noise in phenotypic variation.

Theoretical analysis has indicated that in a fluctuating environment, the heterogeneous population of isogenic cells might achieve a faster growth rate than the homogenous population. An example of this is the petite-positive phenotype (discussed in 2.1.4.) of *S. cerevisiae*. In a medium containing glucose as the sole carbon source, *S. cerevisiae* cells randomly lose mitochondria to produce a heterogeneous population. *S. kluyveri*, a close relative of *S. cerevisiae*, is petite negative, and therefore, the population remains homogenous with respect to mitochondrial distribution during growth on glucose. Interestingly, *S. kluyveri* has a lower growth rate than *S. cerevisiae* when grown in glucose as the sole carbon source, while other growth parameters remain the same. Based on this, it can be conjectured that the increased growth rate of *S. cerevisiae* as compared to *S. kluyveri* in glucose could be due to the petite-positive phenotype, which seems to arise because of stochastic loss of mitochoindria. It is pertinent to note that petite-positive phenotype of *S. cerevisiae* has been hypothesized to be due to the genome duplication followed by massive rearrangement and deletion of unwanted genes. Therefore, it appears that evolution has selected stochasticity for increasing the growth rate.

Long-term adaptation provides yet another interesting example of stochasticity arising out of fluctuation in the distribution of regulatory molecules. A defect in galactose utilization because of a mutation in GAL3 is overcome due to the presence of the degenerate signal transducer *GAL1* (*GAL3* and *GAL1* are paralogues). However, the GAL1p-dependent degenerate pathway in itself is unable to transmit the signal, had it not been for the fragility of the degenerate pathway. Here, the Gal1p-dependent degenerate signaling loop is activated by a positive feed-forward mechanism due to a stochastic decrease in concentration of Gal80p, the repressor, below a threshold, which occurs only in a small fraction of the cell population.

In this example, mutations in the Gal80p can alter its kinetics of interaction with Gal3p, such that the stochastic process (that is turning on of the *GAL* genes through the degenerate signaling pathway) can either be abolished totally or can be increased. That is, evolution can, in principle, fine-tune the outcome of stochasticity. It has been recently demonstrated that a negative feedback through Gal80p (see Chap. 8.2) increases the stochastic transitions to ensure that cells are not trapped in a few phenotypic states but are more free to scan the wide spectrum of environmental conditions. In higher eukaryotes, the stochasticity can be considered even at the level of an individual. For example, in monogenic diseases, individuals bearing the same defective allele exhibit a different degree of phenotypic effects. In addition to the varying genetic background as the cause for such differences, a stochastic process could as well contribute to these differences.

Thus, genome response does not appear to be precisely programmed as we are made to believe But its function is dynamic within the confines of robustness and fragility which brings out deterministic output with utmost precision. Stochasticity seems to provide yet another parameter that evolution can tinker with to increase the evolutionary success of an organisms Unlike the phenotypic variation due to mutations, the phenotypic variation due to stochasticity is more versatile and reversible. By introducing a strong stochastic component into the original framework of the "genetic program" one is able to appreciate better the "ontogenic evolution" (Fig. 8.5.1).

Yeast and humans are evolutionarily separated by 10^9 years. Humans have evolved sophisticated mechanisms to meet the demands of complexity. For example, in support of multicellularity, a complex circulatory system has evolved to provide nutrients efficiently to all cells. In contrast, yeast directly takes up the nutrients from the environment. Yeast overcomes continued nutritional deprivation by switching on a developmental program that converts the normal metabolically

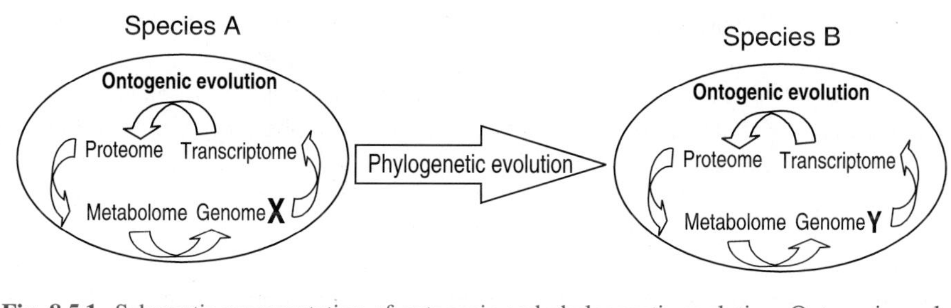

Fig. 8.5.1 Schematic representation of ontogenic and phylogenetic evolution: Ontogenic evolution is dictated by a robust genetic program (X or Y) with a strong component of stochasticity. Despite this, if the organism is unable to respond to the changing conditions fruitfully, then the environment picks up variants that exist in genome X resulting in speciation

active cells into dormant cells referred to as spores. These spores can remain dormant for a long period of time and resume normal metabolic activity immediately upon receiving nutrition. Unlike yeast, humans are vulnerable or fragile to continued nutrient deprivation. The inability to adapt to impending life-threatening situations is the price humans seem to have paid for achieving biological complexity. Evolution only optimizes the living strategy within a context and need not necessarily be progressive. Accordingly yeast and humans seem to have opted different evolutionary paths to meet the demands of environmental vagaries.

References

Balaban NQ, Merrin J, Chait R, Kowalik L, Lebier S (2004) Bacterial persistence as a phenotypic switch. Science 305:1622–1625

Elowitz MB et al. (2002) Stochastic gene expression is a single cell. Science 297:1183–1186

Koern M, Elston TC, Blake WJ, Collins JJ (5005) Stochasticity in gene expression: from theories to phenotypes. Nature 6:451–464

Kemkemer R et al. (2002) Increased noise as an effect of haploinsufficiency of the tumor suppressor gene neurofibromatosis type1 in vitro. Proc Natl Acad Sci USA 99:13783–13788

Kupiec JJ (1997) Darwinian theory for the origin of cellular differentiation. Mol Gen 255:201–208

Hume DA (2000) Probability in transcriptional regulation and its implication for leukocyte differentiation and inducible gene expression Blood 96:2323–2328

McAdaams HH, Arkin A (1999) It is a noisy business. Trends Genet 15:65–69

Thattai M, van Oudenaaredn A (2004) Stochastic gene expression in fluctuating environment. Genetics 167:523–520

Gally AJ, Edelman GM (2001) Degeneracy and complexity in biological systems. Proc Nat Acad Sci USA 98:13763–13768

Raser MJ, O'Shea EK (2005) Noise in gene expression: Origins, consequence and control. Science 309:2010–2013

Swain PS, Elowitz MB, Siggia ED (2002) Intrinsic and extrinsic contributions to stochasticity in gene expression. Proc Nat Acad Sci USA 99:12795–12800

Stelling J, Sauer U, Szallasi Z, Doyle FJ, Doyle J (2004) Robustness of cellular functions. Cell 118:675–685

Wagner A (2005) Robustness and evolvability in living systems. Princeton University Press, Princeton

Alon U (2007) An introduction to systems biology. Chapman & Hall, New York

Index

Printing: Krips bv, Meppel, The Netherlands
Binding: Stürtz, Würzburg, Germany